The Dyna
Nucl
Proliferation

£ 1.00

The Dynamics of
Nuclear
Proliferation

With a Foreword by Joseph S. Nye, Jr.

Stephen M. Meyer

The University of Chicago Press
Chicago and London

The University of Chicago Press, Chicago 60637
The University of Chicago Press, Ltd., London
© 1984 by The University of Chicago
All rights reserved. Published 1984
Paperback edition 1986
Printed in the United States of America
95 94 93 92 91 90 89 88 87 86   6 5 4 3 2

Library of Congress Cataloging in Publication Data

Meyer, Stephen M.
   The dynamics of nuclear proliferation.

   Revision of thesis (Ph.D.)—University of Michigan, 1978.
   Bibliography: p.
   Includes index.
   1. Atomic weapons. 2. World politics—1945–
I. Title.
U264.M49   1984   355.8′25119   83-17893
ISBN 0-226-52148-6 (cloth)
ISBN 0-226-52149-4 (paper)

To My Parents

Contents

Foreword

JOSEPH S. NYE, JR.

Twenty years ago, John F. Kennedy saw the possibility of a world in the 1970s with fifteen to twenty-five nuclear weapons states, a situation he regarded as "the greatest possible danger." Instead, the seventies closed with five declared weapons states, one state that had launched a "peaceful explosion," and one or two that were believed to be just below the explosion threshold. Given that nuclear weaponry is a forty-year-old technology, what is surprising is not that it has spread, but that it has not spread further.

This is one of the puzzles Stephen Meyer addresses in this fine study of nuclear proliferation. What capabilities and incentives lead states to cross the nuclear threshold and develop nuclear weapons? What has been the experience of states that have done so? Why has the rate of new entrants slowed? Will this pattern continue into the future, or is it likely to change? These questions need clear answers if we are to frame sensible policies. Using a combination of social science and common sense, Meyer helps clarify the problems. He does not stake out a position in the heated debates over nonproliferation policy, but his analysis will be essential for those who do as well as for those who merely seek to understand this major issue.

While it is unlikely that any policy can totally prevent further proliferation, the task of slowing the rate of spread is far from hopeless—as the following pages demonstrate.

If we specify a goal of reducing the rate and degree of proliferation so as to manage its destabilizing effects and reduce the prospects of nuclear use, then there are many promising tasks for nonproliferation policy. Even if one were to accept the sanguine view that nuclear spread may not be destabilizing in all cases, the rate of proliferation affects the likelihood of destabilizing effects. And a sanguine prognosis is far from certain. Sensible policy must hedge against potentially large downside risks.

Proliferation is sometimes conceived in simple terms of a single explosion. Indeed, that concept is enshrined in the Non-Proliferation Treaty (NPT). But it can also be conceptualized as analogous to a staircase with many steps before and after a first nuclear test. A first explosion is politically important as a key landing in the staircase, but militarily a single crude explosive device does not bring entry into some meaningful nuclear "club." The very idea of a nuclear club is misleading. The difference between a single crude device and a modern nuclear arsenal is as stark as the difference between having an apple and having an orchard.

There are various reasons why this is so, including the restrictive policies of the weapons states, the calculated self-interest of many nonweapons states in forgoing nuclear weapons, and the development of an international regime of treaties, rules, and procedures that establishes a general presumption against proliferation. The main norms and practices of this regime are found in the Non-Proliferation Treaty and its regional counterparts trying to work toward a nonnuclear Latin America, in the safeguards, rules, and procedures of the International Atomic Energy Agency (IAEA), and in various United Nations resolutions. While there are a few important exceptions, the large majority of states adhere to at least part of this set of norms.

In the early 1970s there was a degree of complacency about this nonproliferation regime. Such complacency was shattered in 1974 by the Indian explosion of a "peaceful" nuclear device using plutonium derived from a Canadian-supplied research reactor and by the oil crisis, which led to a sudden surge of exaggerated expectations about the importance of nuclear energy, including fears that uranium supplies would be exhausted.

Various nations developed plans for early commercial use of plutonium fuels, and some countries such as Korea and Pakistan arranged to import allegedly commercial processing plants for what later were disclosed to be nuclear explosives programs. Typical projections for the 1980s from this period saw eight reactors in Bangladesh and thirty to forty countries using plutonium fuels by the end of the decade. With this challenge to the regime, it is not surprising that policy responses in the late 1970s such as the London Suppliers Group and International Nuclear Fuel Cycle Evaluation (INFCE) focused on general questions of capabilities.

Some of the heated debates of the 1970s about nuclear energy issues and their relation to proliferation have now been answered by experience, economic trends, and the conclusions of INFCE. For example, the argument that no state would find it rational to try to misuse a "peaceful" nuclear facility rather than build a dedicated weapons facility has been disproved by experience. The view that shortages of uranium would require early use of plutonium has succumbed to more realistic economic projections. The view that nuclear energy would provide energy security has also been belied by reality and better analysis: the great threats to energy security in the 1980s are from political interruptions of Persian Gulf oil, for which the appropriate answers are free markets backed by emergency stockpiles, conservation, and coal, not nuclear plants with ten-year lead times. And the optimistic projections about nuclear energy in developing countries have also had to be toned down for most countries in the light of economic experience. Of sixteen less developed countries currently operating reactors, only half a dozen have significant power programs.

The nonproliferation problems of the 1980s are more likely to have a political cast and to require political solutions. This does not mean nuclear energy and capability issues will not be important, but they are only one of the major problems involved in managing proliferation in the coming decade. Political issues related to security guarantees, sanctions, and devices to maintain the current nonproliferation regime will also be critical. Balancing the problems of capabilities and intentions will remain at the heart of the policy problem. And nowhere are the facts of past experience combined with sensitive projections about the future in a more thorough and dispassionate way than they are in this book. Policy wisdom begins with analytic clarity and insight. That is what Meyer provides in the pages that follow.

Preface

Among those with special interest in the proliferation of nuclear weapons states, there is widespread agreement that further proliferation will pose increased threats to international peace and security. Correspondingly, research writing and policy in this area have generally emphasized strategies to halt, restrain, and control further nuclear proliferation. Prescriptions have ranged from simple pledges of nuclear abstinence to grand schemes for global nuclear control regimes. Some have argued strenuously for the control of nuclear technology, while others have called for conflict resolution and general disarmament.

Underlying these prescriptions for lessening the rate and scope of future nuclear proliferation are explicit and implicit assumptions about the "forces" that drive the nuclear proliferation process and about other factors that may influence its course. Some place the responsibility for the spread of nuclear weapons on technological forces, while others stress politico-military forces. Even within each group, there are substantial differences in opinion regarding the relative weights to assign to various components.

Although the debate over the wisdom of policy options and alternatives has continued for almost two decades, there has yet to be a systematic examination of the empirical validity of the assumptions underlying the various nonproliferation policy prescriptions. Each school of thought is already convinced of the "obvious" empirical validity of its own view and the obvious naiveté and simplemindedness—if not ignorance—of the views of others. Moreover, it is often asserted that no one of consequence really subscribes to any of the "other" contending views. Therefore any systematic test of assumptions and hypotheses is perceived as a waste of effort: it will confirm the obvious while finding nothing of value in examining "straw men" (other schools of thought). Of course the problem is that each of the contending schools feels this way.

I must reject such blanket appeals to faith. There seems to be little sense in haggling over which combination of technological controls, diplomatic, military, and political activities, and international cooperation programs is likely to minimize future nuclear weapons proliferation until we have at least tried to establish the empirical validity of what are currently little more than impressions of the proliferation process. Thus the objective of this book is to take one step back—ignoring the policy disputes—and undertake a rigorous and systematic examination of the assumptions and contending hypotheses that constitute contemporary thinking on nuclear proliferation. The intent is not to determine who is right or who is wrong, for it is likely that some of the assumptions of each school are valid. Rather, the intent is to develop a better picture by using the various schools of thought as analytic windows. A more precise understanding of how the process operates should offer better guidance for predicting future nuclear proliferation and, ultimately, for controlling it. Indeed, my hope is that, in the future, there will be no additional empirical data to analyze.

Acknowledgments

There are many individuals and organizations to whom I am indebted. Initial research for this study began more than a half dozen years ago at the University of Michigan and benefited greatly from the advice of Harold Jacobson, Catherine Kelleher, and David Singer. More recently, Albert Carnesale (Harvard University), Ted Greenwood (MIT), Michael Intriligator (UCLA), and William Potter (UCLA) have provided valuable comments and criticisms on earlier drafts of this manuscript.

Research support throughout this project has been provided by the Institute for the Study of World Politics. Most notably, their generous assistance financed the extensive data making and computer analysis that constitute the body of the project.

This book was written at the Center for International Studies, Massachusetts Institute of Technology, and I am grateful to the Center for its support. Toni Paganis and Chuck Lockman typed innumerable drafts of the manuscript with care and efficiency.

1
Nuclear Proliferation: Contending Views

The most widely held image of nuclear proliferation in progress is a country sedulously working to build and test an atomic bomb. The traditional indicator of nuclear proliferation is its first nuclear test explosion, marking the birth of the nth nuclear weapons power. To be sure, case histories of all five recognized nuclear weapons powers faithfully reflect this view of the nuclear proliferation process. All five nations doggedly pursued nuclear technology explicitly for the purpose of producing weapons. Through the late 1950s, the capability to manufacture nuclear weapons was largely the result of concerted efforts to do just that.

But there are compelling reasons to question whether this will continue to be the most common form of proliferation in the future. For a long time the "nuclear" aspect tended to obscure the fact that the initial process of manufacturing nuclear weapons is largely an exercise in chemical, mechanical, and industrial engineering. The release of vast quantities of nuclear science and engineering data through the Atoms for Peace program and its successors removed many "unconventional" technical hurdles. Thus, one result of the global push for scientific, technical, industrial, and economic development over the past several decades is that many countries have acquired the basic skills and resources necessary for manufacturing nuclear weapons. In other words, irrespective of specific government interest—or lack of interest—in "going nuclear," a number of countries now possess *latent capacities* to produce nuclear weaponry.[1] Capability has become decoupled from prior interest in building nuclear weapons.

Further accentuating the break with the past was the growth of civil nuclear programs. First there were nuclear science research

1

programs, usually built around a research reactor and several laboratories. As is shown in figure 1, the growth in the number of countries with at least one research reactor during the first decade of

Fig. 1. Cumulative number of countries with research reactor facilities.

the nuclear age was relatively slow. Between 1945 and 1955 only four countries were operating research reactors. In a sense this reflects the prevailing juncture of indigenous capability and interest in nuclear science research. However, as the Atoms for Peace program got into full swing, the great research reactor "giveaway" began. Between 1955 and 1965, thirty-nine additional countries acquired reactors—mostly as gifts from the United States and the Soviet Union. It is unclear whether this explosion of interest in possessing research reactors was due to demand pull or supply push, but in either case the accessibility of the tools for nuclear research increased vastly during the second decade of the nuclear age. Moreover, extensive international training programs were conducted in the United States, Britain, France, and the Soviet Union.

By the time the increase in nations operating research reactors began to tail off, interest and investment in civil nuclear power facilities was widespread. As is seen in figure 2, during the second decade of the nuclear age six nations began to operate nuclear power reactors. By the end of the third decade, with its promise of cheap electrical energy, thirteen more nations had nuclear power reactors. And between 1975 and 1982 eight additional states obtained nuclear

Fig. 2. Cumulative number of countries with nuclear power facilities.

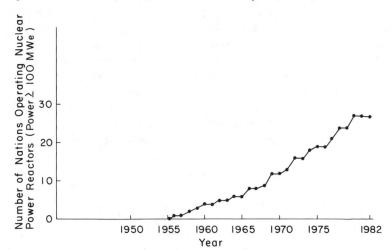

power reactors. Slowly but surely, countries began to develop substantial nuclear infrastructures: personnel with the requisite knowledge and skills, and equipment and plants with productive capacity that could be systematically converted to nuclear weapons production. The link between indigenous nuclear development and nuclear weapons ambitions was broken. Thus, while only six countries had tested nuclear explosive devices by the end of 1982, another thirty-five or so possessed latent capacities to do so, and many of them had well-developed nuclear infrastructures.

As more and more countries acquired latent capacities supported by significant nuclear infrastructures, students of nuclear proliferation became concerned about the implications of this creeping evolution. How would it affect the "cost-benefit" calculus and the "lag times" of "going nuclear," the perception of threat among adversaries, and the prospects for early detection of the appearance of new nuclear weapons countries?[2] Nuclear proliferation could no longer be viewed as the well-defined black-to-white jump to nuclear weapons status. Instead, it had to be seen as a developmental process reflecting the growth of latent capacities around the globe. That between 1964 and 1974 no other countries besides the five recognized nuclear weapons powers had detonated a nuclear weapon did not mean nuclear proliferation had not advanced. On the contrary, it was widely perceived as moving forward in the form of ever-developing latent capacities encompassing substantial nuclear infrastructures.

3

The research and policy communities increasingly addressed the problem in terms of the global creep of nuclear technology and hardware. Studies by the Office of Technology Assessment (1977), the Department of Energy (and its organizational predecessors), the FORD-MITRE group (1977), and NASAP (1980) all tried to assess technological dimensions of nuclear proliferation in the 1980s and to explore its implications and United States policy options. The London Suppliers Group of nuclear technology exporters gathered to try to agree upon a set of rules for marketing nuclear technology. The hope (at least from the perspective of those most concerned about nuclear proliferation) was to make it more difficult for prospective nth countries to acquire sensitive nuclear technology, hardware, and materials. And as concern about the proliferation implications of plutonium recycling mounted, the United States government initiated the International Fuel Cycle Evaluation (INFCE) conference. Among its many goals, it hoped that INFCE first would convince skeptical nations that the impending move to plutonium recycling would indeed exacerbate nuclear proliferation. Then, it hoped, the INFCE working groups would validate United States government claims that near-term plutonium recycling was both uneconomical and unnecessary (since plenty of exploitable nuclear fuel was available).

Unfortunately, this strong emphasis on technical aspects led large segments of the academic and policy communities to perceive nuclear weapons proliferation as almost entirely a technological problem—much like air pollution or toxic waste disposal. "Intentions can always change, so we should concentrate on capabilities." Many came to see every national effort to add to indigenous nuclear infrastructures as nuclear proliferation, and any decline as a reduction of nuclear proliferation. The German-Brazilian nuclear-technology agreement consequently was considered an advance in nuclear proliferation, while the cancelation of the French-Pakistani deal to provide Pakistan with reprocessing technology was viewed as a retreat, if not a defeat. This thinking also gave rise to an emphasis on the descriptive dimensions of nuclear proliferation: the number of nations with latent capacities, changes in the "time to bomb" options of prospective nth countries, the rate of power reactor purchases, and growth in global plutonium stocks. How many countries would be within two years of "the bomb" by 1990? How many twenty-kiloton weapons could be fashioned out of global plutonium stocks in 1985? What would it cost the Vatican to build "the Catholic" bomb in 1987?

But these all beg the real question. Consider, for example, the fact that the globe is covered with insecticide production plants that, at a cost of several thousands of dollars, could be rapidly converted to manufacture deadly nerve gas—enough to wipe out entire countries. Yet no one seems greatly concerned, probably because there is little reason to believe few if any countries have even considered this. But this is precisely the crux of the nuclear proliferation issue: nations do *decide* to exercise their option and "go nuclear." The various technical, political, and military effects of the forward creep of latent capacities are significant only because there is reason to believe that some nations will decide to transform their latent capacities into operational capabilities. In particular, latent capacities and nuclear infrastructures are important because of the ways they affect government decisions to jump to operational capabilities. How does the state of a nation's latent capacity affect the likelihood of its pursuing nuclear weapons production? Does having an advanced nuclear infrastructure make a decision to go nuclear more likely? Indeed, what pushes a government to decide to make the jump from latent capacity to operational capability? Why do governments decide the time has come for their countries to "go nuclear"?

DECISIONS TO INITIATE NUCLEAR WEAPONS PROGRAMS

In pursuing this question of *why* nations "go nuclear," the pivotal point in the nuclear proliferation process is the decision to pursue nuclear weapons acquisition—not having the first weapon actually in hand. Distinctions between a capability decision, acquiring (or possessing) a latent capacity, a proliferation decision, and ultimately possessing functional nuclear weapons are crucial to understanding the nuclear proliferation process. Yet they are often glossed over or altogether ignored. As I noted, nations may acquire the fundamental capability to produce nuclear weapons by intentional effort or as an unintended by-product of industrial and economic development. In the former case, they make a *capability decision*—an explicit government decision to develop a latent capacity that provides an indigenous capability to implement and support a nuclear weapons program. A capability decision may occur before or in conjuncion with a proliferation decision. A capability decision in the absence of a proliferation decision reflects developing "a nuclear option," enhancing a nuclear option, and keeping a nuclear option open. The Swedish government is one of several believed to have followed this path. Through explicit effort (a capability decision), Sweden acquired a latent capacity long ago, yet no Swedish nuclear weapons were ever developed for lack of a proliferation decision.[3] In contrast, the Bhutto government of Pakistan is reported to have made simultaneous proliferation and capability decisions in the early 1970s. Pakistan set about developing a latent capacity first by attempting to import a commercial reprocessing facility, then by covertly acquiring uranium-enrichment technology. With its latent capacity mostly in place and its proliferation decision still in force, Pakistan is attempting to produce its first nuclear weapons (Dunn 1982, 31). Thus it is obvious that capability decisions in isolation—keeping the nuclear option open—are not equivalent to capability decisions that follow from proliferation decisions to "get the bomb."

Regardless of how a country acquires a latent capacity, the capstone of the nuclear proliferation process is the acquisition of functional nuclear weapons, something that could come about only from an explicit government decision—a proliferation decision—to transform a latent capacity into an operational capability. Nuclear weapons do not generate spontaneously from stockpiles of fissile material. Thus the decision to "go nuclear" is the crucial step in the nuclear proliferation process.

A proliferation decision may or may not lead ultimately to nuclear weapons. Real-world problems may alter or undermine planned nuclear weapons production timetables. Countries attempting to manufacture nuclear weapons may find that the technical hurdles are greater than expected. Other countries' efforts to hinder the prospective proliferant's program (e.g., Israel's attack on the Iraqi nuclear facility, or United States diplomatic intervention) must be considered as well. Thus, while we can expect that decisions to begin a nuclear weapons program will be made about the time the "conditions" that prompt them are manifest, the same cannot be assumed for the physical acquisition of the weapons. Moreover, this emphasis on decision rather than outcome also implies that failure to produce functional nuclear weapons should not be allowed to mask the fact that proliferation decisions indeed were made. Decisions to pursue nuclear weapons development made by Nazi Germany (1940) and Imperial Japan (1941) are appropriate examples. A more recent example is South Korea (1972). For reasons I will discuss in the next chapter, we might also include the aborted Indian program to produce a peaceful nuclear explosive (PNE) device in 1965.

Of course nations may attempt to acquire nuclear weapons by more unorthodox methods—theft or purchase. But even here, where the objective is to forgo indigenous nuclear weapons production, prospective proliferants may discover, much to their dismay, that the security of others' nuclear weapons is better than they had believed or that there are no willing sellers. There have been reports that in 1969 Libya's Colonel Muammar Qaddafi offered the Chinese $1 billion for an atomic bomb, and that a recent second effort resulted in his losing $100,000 to an enterprising confidence man. Here too failure to acquire "the bomb" should not obscure the fact that the decision and subsequent effort were made.[4]

In general, proliferation decisions that precede capability decisions, and those that look to theft or purchase as nuclear weapons acquisition strategies, are almost impossible to pin down a priori. There is no way to define this subpopulation of decisions systematically because there are so few telltale signs. While it is possible to catalog anecdotes and rumors, there may be many governments that are eager to buy or steal atomic bombs yet do nothing because they perceive the likelihood as too small. But if we could develop a systematic understanding of why nations go nuclear—or at least devise a set of reliable indicators of *decisions* to do so—it might be possible to determine which countries are most likely to try to acquire nuclear weapons by theft or purchase. Thus, for reasons of policy interest (concern about the implications of growing nuclear infrastructures) and for the reasons of research design outlined above, it is necessary to begin this examination of "why nations go nuclear" with a look at those countries that already possess the basic capability to go nuclear by indigenous effort to produce nuclear weapons. Why do governments decide to transform existing latent capacities into operational capabilities? Why do nations initiate nuclear weapons programs? Afterward we can broaden our examination to proliferation decisions (or the lack thereof) where no latent capacity exists and where nuclear weapons acquisition includes theft and purchase.

From this perspective, table 1 lists thirteen historical decisions to initiate nuclear weapons programs where latent capacities already existed. The first five—Germany, Japan, the United States, the United Kingdom (1), and the Soviet Union—were wartime programs, undertaken with no certainty that nuclear weapons could actually be built. The next three—the United Kingdom (2), France, and the People's Republic of China—were of cold war vintage. And the last five—India (1 & 2), Israel, South Africa, and the Republic of Korea—are what might be most similar to nuclear proliferation as it will appear in the future: non-European, Third World nuclear weapons efforts. (All these cases are discussed in greater detail in Appendix A.) To reiterate, each of these proliferation decisions occurred in the presence of existing latent capacities—that is, each of the countries possessed the fundamental industrial, scientific,

TABLE I
DATES OF DECISIONS TO INITIATE NUCLEAR WEAPONS PROGRAMS

Country	Year
Nazi Germany	1940[a]
Imperial Japan	1941[a]
United States	1942
United Kingdom (1)	1942
Soviet Union	1942
United Kingdom (2)	1947
France	1956
People's Republic of China	1957
India (1)	1965
Israel	1968
India (2)	1972
Republic of Korea ·	1972
Republic of South Africa	1975

[a]These are uncompleted programs that involved scientific feasibility studies.

engineering, and financial resources required to pursue an indigenous dedicated program for nuclear weapons production.

Notably missing from the list are the alleged cases of Argentina (1950s), Indonesia (1960s), Taiwan (1974), Pakistan (1972), and Libya (1969 to the present). Under the Peron regime in the early 1950s, Argentina embarked on a bizarre course to develop nuclear fusion. The project, led by an Austrian physicist, was simply a hoax designed to get money from the Argentine government. In the Taiwan instance, there is insufficient case study evidence of a decision to pursue nuclear weapons manufacture. Most accounts suggest that the Taiwanese government was trying to advance its nuclear option short of producing functional weapons—it was implementing a capability decision toward developing a more sophisticated latent capacity.

Pakistan and Libya fall outside the subset of countries with preexisting latent capacities that we are now examining—at least up to 1982. However, these two cases—and others like Iraq—are considered in later chapters when the latent capacity criterion is relaxed.

WHY NATIONS GO NUCLEAR: CONTENDING HYPOTHESES

Although substantial amounts of work have gone into describing and analyzing the many facets of the technological aspect of nuclear proliferation, analyses of decision making related to the acquisition of operational capabilities are few and far between. To be sure, there are numerous excellent historical case studies that examine nuclear decision making in individual countries (Yaeger 1980; Dunn and Kahn 1976; Lefever 1978; Dunn 1982). However, what is lacking are more general studies of nuclear decision making as the subject of analysis itself. Is there a systematic pattern that underlies decisions to acquire nuclear weapons? The few efforts one can identify as leaning in this direction tend to assert relationships and causality rather than test for them. While such works are illuminating and thought provoking, they nonetheless rest on untested hypotheses and assumptions.

As it turns out, three general classes of hypotheses—or schools of thought—can be identified in the nuclear proliferation literature. The first class posits that nuclear technology itself is the driving force behind decisions to acquire nuclear weapons—that a technology imperative pushes nations from latent capacity to operational capability. Governments "decide" to go nuclear because the technology is available, thereby making the technical/financial costs manageable and the opportunity irresistible.[5] The second class of hypotheses sees the quest for nuclear weaponry as resulting from the systematic effects of a discrete set of political and military variables. Nuclear weapons are one of a number of policy options nations may pursue in trying to accomplish foreign, defense, and domestic policy objectives. Proliferation decisions therefore are motivated by political and military considerations, and when the proper political-military conditions come together a proliferation decision follows. The third class of hypotheses views the nuclear proliferation process as largely idiographic. Countries "go nuclear" because particular individuals and particular events come together at specific times and create the proper conditions. However, the mixing of variables is random and yields unpredictable results. Thus decisions to initiate nuclear weapons programs are sui generis.

THE TECHNOLOGICAL IMPERATIVE: I

While there is no formal technological imperative model, there are a string of notions that may be reduced to a single maxim: once a nation acquires the capability to manufacture nuclear weaponry (has latent capacity), it will inevitably do so (move to operational capability). The deterministic nature of this kind of hypothesis is readily apparent when working premises are enumerated:

1. As the result of continuous national effort to improve the level of economic development, the underlying industrial/technological capacity of the nation will progress.

9

2. Once the production of nuclear weapons becomes techno-logically and industrially feasible, the sheer momentum of technological progress, coupled with the challenge of turning an idea into a product, will compel the nation to complete the process and "go nuclear."
3. Consequently, all nations will eventually cross the threshold from a latent capacity to the active manufacture of nuclear weapons.

As Dr. Herbert York remarked in reference to the reports of a secret Japanese atomic weapons program during World War II:

> the Japanese story completes the set, that every nation that might plausibly have started nuclear weapons programs did so: Germany, Great Britain, the United States, the Soviet Union, France, and we now know, Japan. So the case has been weakened for those who have argued that governments, or more precisely, generals, emperors, and presidents can hold back from the decision and say "No." The decision to develop nuclear weapons is not a fluke of certain governments, but a general technological imperative.[6]

Where York was referring to past events, energy specialist Amory Lovins (1980, 1138, 1176) sees nuclear technology as "the main driving force" behind nuclear proliferation and "its greatest cause" in the future. Similarly, a number of senators and congressmen could be heard echoing this view in the wake of Israel's attack on Iraq's Osiraq research reactor in June 1981. Then, too, if one presumes that the incentives to acquire nuclear weapons are ever present—that all countries would like to have nuclear weapons—then the only determining factor becomes technology.

Certainly the descriptive and analytic simplicity of this model of the technological imperative hypothesis must be appreciated. Following a strict interpretation, it obviates the need to give serious attention to the domestic and international milieu (the situation) or to any other contextual factors *not* directly related to the national resource capacity.[7] Countries simply "go nuclear" within some period after acquiring a latent capacity, if not as soon as they are able. Of course this model does recognize that humans must still make the relevant decisions. But the argument is that the technological momentum is so strong an influence, and the demand for nuclear weapons is so pervasive, that decision makers are "carried along." They proceed simply by the logic of not leaving a promising technological path untraveled.

Technological Imperative: II

Yet despite its simplicity and elegance the strict version of the technological imperative hypothesis described above is not historically compelling, at least as it applies to nuclear proliferation. In particular, it is hard to explain the nonnuclear weapons status of nations like Argentina, Brazil, Spain, Australia, Belgium, Canada, Italy, the Netherlands, Sweden, and Switzerland. To be sure, these are all countries "that might plausibly have started nuclear weapons programs." But, in contrast to what the strict technological imperative model would lead us to expect to observe, they have not gone on to manufacture nuclear weapons.

Perhaps a more realistic interpretation of the technological imperative hypothesis would postulate that the overall manifestation of the technological imperative is spread out randomly over time. In essence, though some nations may take longer than others to respond to the technological imperative, all will eventually succumb. In this view latent capacities are merely accidents looking for a place to happen, and eventually the irresistibility of nuclear weapons wins out. True, nations like Argentina, Brazil, and Sweden have not yet crossed the threshold, but they will in time. By analogy, J. Robert Oppenheimer, speaking of the United States decision to proceed with thermonuclear weapons development, observed, "it was so technically sweet, we had to do it" (quoted in Lefever 1979, 21). This model is perhaps closer to York's own view as well.

The implication is that proliferation decisions will be distributed across time in a manner *not* systematically related to the simple attainment of a latent capacity. Thus this "delayed" version of the technological imperative hypothesis does not directly link the decision to "go nuclear" with the attainment of the capability to do so other than to stipulate that the former will inevitably follow the latter. Here too the particular "reasons" why decision makers believe they have decided to pursue nuclear weapons production are seen as peripheral. Possessing a latent capacity to produce nuclear weapons so warps the decision process that nuclear weapons eventually end up as the "right answer" to some policy problem. While the timing may be probabilistic, the decision itself is inevitable.

Technological Imperative: III

A third variation of the technological imperative—and one that perhaps better reflects Lovins's view—argues that the greater the level of nuclear-related infrastructure in a country, the more likely it is to "go nuclear." Here the driving force of nuclear technology is postulated as being directly proportional to the level of development of a nation's latent capacity. Thus Egypt with power reactors would be more likely to pursue nuclear weapons production than Egypt without them.

Actually, this is just a more sophisticated version of the technological imperative II hypothesis. Where version II postulated a "random" distribution of decisions over time, this version argues for a systematic pattern tied to the prevailing level of nuclear infrastructure. As the level of nuclear infrastructure rises, the expected burden of a nuclear weapons program declines. Moreover, even nonresource "costs" may be perceived as less weighty—that is, the technical ease of producing nuclear weapons may affect decision making psychology by artificially inflating perceived benefits and deflating perceived costs. So the decision process is warped in proportion to the ease with which nuclear weapons can be produced.

And so we end up with three models of the technological imperative hypothesis. Each argues—albeit with its own twist—that a latent capacity to manufacture nuclear weapon is, in itself, sufficient to produce a decision to do so. Simply stated: all nations with latent capacities will eventually acquire nuclear weapons. In this respect the notion that some technological imperative drives the nuclear proliferation process should not be seen as a "straw man." There are some interesting parallels between technological imperative arguments and the ongoing gun-control debate in the United States. Do guns kill people or do people kill people? Those who are against gun control argue that people kill people and that the gun is merely the chosen instrument. Without guns people would use something else. Those who favor gun control argue that guns do kill people in that guns make killing so easy and "accessible" that a wider range of disputes or disagreements serve as an "excuse" to kill. Capabilities are seen as affecting psychology and decision making. In a similar fashion, the notion of a technological imperative argues that the capability to produce nuclear weapons makes producing "the bomb" so effortless that practically any policy problem becomes a logical reason to acquire nuclear weapons.

OPTIONS AND CHOICE: THE MOTIVATIONAL ASPECT OF NUCLEAR PROLIFERATION

In contrast to the long-run determinism of the technological imperative, the motivational hypothesis involves an inherently *probabilis-*

tic process. Rather than being driven by some irresistible force, nations—or, more precisely, their governments—are posited as having the choice of whether to launch a nuclear weapons program. In this respect the motivational hypothesis reflects a distinctly different notion about how nations behave.

Technical capability is cast in the role of a *necessary condition* for nuclear proliferation but not a sufficient condition. But just because a nuclear weapons program seems feasible (the nation possesses, or could acquire, a latent capacity) does not imply that a decision to begin such a program will be made. Instead, it is specific politico-military conditions that motivate—that is, cause—national decision makers to initiate a nuclear weapons program. It is the convergence of a latent capacity (the first necessary condition) with significant proliferation-related motivations (the second necessary condition) that results in nuclear proliferation.[8]

Here the motivational model brings together the concepts of *decision stimuli, decision options, and choice.* Given that the acquisition of nuclear weapons is a matter of "high politics," for the moment it will be convenient to assume that national decisions are made (though not necessarily implemented) by a single discrete decision unit ("the government"). Accordingly, *decision stimuli* connotes conditions, or changes in conditions (internal and/or external to the decision unit), that evoke the need for the decision unit to respond. Decision stimuli may be either long-standing conditions or discrete events. One possible decision stimulus might be an existing military rivalry in which the adversary has over-whelming conventional superiority. The *decision options*, then, would be the set of responses the decision makers perceive as available. For a nonnuclear weapons country faced with an over-whelming conventional threat these might include ignoring the situation, increasing conventional military capabilities, searching for security guarantees from other countries, or "going nuclear." *Choice*, obviously, signifies the selection of a specific decision option.

Another decision stimulus might be the knowledge that an adversary nation possesses a latent capacity. Depending upon how this "potential threat" is interpreted, the response might be selected from among the following decision options: initiate a nuclear weapons program, increase basic scientific and engineering research on nuclear weaponry, increase covert intelligence activities while taking a wait-and-see stance, or ignore the information. Of course a nation might pursue several options simultaneously.

The ability to convert a latent capacity into an operational capability is, in every sense of the term, a decision option. It represents a distinct course of action that a government can independently implement in response to certain appropriately perceived stimuli (e.g., a perceived increase in military threats). Hence the term "nuclear option" as it appears in the literature is right on the mark. According to the motivational hypothesis, a nation is expected to exercise its *nuclear option*, transforming its latent capacity into an operational capability should conditions warrant it.

Correspondingly, let us define the collection of specific conditions, or changes in conditions, that prompt explicit consideration of the nuclear option as *motive factors*, or *motive conditions*, for acquiring nuclear weapons. In an analogous fashion, there are specific conditions, or changes in conditions, that act to "devalue" the nuclear option. These may be termed *dissuasive factors*, or *dissuasive conditions*. Exactly what these proliferation-related motive and dissuasive factors are will be discussed in detail somewhat later. In any case, taken together these two groups of conditions form the *motivational basis of nuclear proliferation*.

Now, once the nuclear option has been raised, its relative attractiveness, compared with the other possible decision options, will be determined by the perceived mix of *incentives and disincentives* attached to each particular option. In economic terminology, the relative benefits (incentives) and costs (disincentives) of each option will be compared and weighed.[9] Looked at in a slightly different manner, motive factors for acquiring nuclear weapons reflect existing conditions (or ongoing changes in conditions) that prompt consideration of the nuclear option by the decision unit. The incentives for "going nuclear," then, become the desired conditions perceived as being attainable through the national acquisition of nuclear weaponry, while the disincentives become the undesirable (dissuasive) conditions (from the persective of the decision makers) likely to be created. The relative strengths of the incentives for, and the disincentives against, a specific nation's going nuclear will therefore depend on the particular conditions (both motive and dissuasive factors) that are manifest when the nuclear option is first considered. This formulation also includes perceptions of future conditions. Clearly, then, the relative and absolute strengths of proliferation-related incentives and disincentives cannot be characterized as fixed quantities (or as stable forces). On the contrary, being condition dependent, they should be expected to vary across both time and space.

By way of illustration, let us consider the present case of South Korea. Those who subscribe to the motivational hypothesis point out that South Korea's primary motivation for acquiring atomic weapons would be the conventional military threat posed by North Korea and North Korea's allies. Thus some incentives for South Korea to exercise its nuclear option are the perceived possibilities for enhancing its overall military capability and using the fear of nuclear escalation to deter a North Korean attack. Relevant dissuasive conditions might include the present deployment of United States troops in South Korea, the accompanying United States tactical nuclear forces, and South Korea's ratification of the Non-Proliferation Treaty (NPT). The corresponding disincentives would be the possible United States response of withdrawing its troops and tactical nuclear forces stationed in South Korea, and a most probable adverse international reaction to the abrogation of the NPT.

As further illustrations, consider Sweden and the Republic of South Africa. One disincentive they share is the adverse political reaction of other nations. But there can be little doubt that the strength of this disincentive is greater for Sweden, situated among friendly countries and with an international image of high repute, than it is for South Africa, which is bordered by hostile nations and considered an international pariah. The point is simple: though the same disincentives (or incentives) may operate across time and space, their influence on the choice between options may vary considerably—presumably as a function of the combined effects of other attendant conditions.

Finally, we have the notions of *aids* to nuclear proliferation and *restraints* on nuclear proliferation. Aids facilitate a country's drive to manufacture atomic weapons—they make the physical task of going nuclear easier. Restraints hinder the acquisition of atomic weapons. Aids and restraints differ from incentives and disincentives in that the former are objective and tangible items that refer to technical capabilities, while the latter are subjectively perceived cognitions associated with proliferation-related motivations. (Unfortunately, these concepts are often used interchangeably in the literature, which has blurred this very important distinction.) An aid to nuclear proliferation would be an offer by the United States to sell atomic bombs at ten dollars a kiloton. Under the motivational hypothesis, a national civil nuclear power program of reactors, reprocessing plants, and/or enrichment facilities is seen as an aid to nuclear weapons proliferation in that it would make "going nuclear" easier. But it is not an incentive to go nuclear.

A restraint on nuclear proliferation is the limitation on sensitive nuclear technology exports imposed by the London Suppliers

Group. Diversion-resistant nuclear technologies also serve as restraints, as do some forms of nuclear safeguards.

Obviously, aids and constraints may act as catalysts in proliferation decision making. The potency of a given mix of incentives and disincentives may be affected by aids and constraints. But aids and constraints alone—in the absence of motive conditions—will not affect proliferation decision making. Up to this point I have not mentioned the assumptions the motivational hypothesis makes regarding the nature of government decision making (e.g., rationality). In fact, all that needs to be assumed is that (1) the decision unit is a discrete entity in the sense that where there are several key individuals within a decision unit (e.g., the Soviet Politburo) they all have roughly the same general views about the world (though the group's composition and structure may change over time); (2) the decision unit exists within the context of a defined situation (or set of conditions) that may change over time; (3) decision makers do perceive "their" situation and its manifest changes; (4) decision makers may react to certain conditions or changes in conditions; and (5) there is usually a range of alternative responses available.

As it turns out, the decision rules delineated above are general enough to be subsumed under any number of decision making paradigms. Whether one considers the rational actor (analytic) model (Snyder, Bruck, and Sapin 1962), the bureaucratic model (Halperin 1974; Allison 1979), the organizational model (Cyert and March 1963), the cybernetic model (Dutsch 1963; Steinbruner 1974), or any one of a number of psychological models (DeRivera 1968; Janis 1972), the notions of stimuli and responses are common to all. Where they differ is in their respective assumptions about how stimuli and responses are linked (i.e., how information is perceived, how it is "processed," and the algorithms by which choices are made). However, as I outlined above, the motivational model does not require any assumptions about *how* the decision process operates. All it postulates is that certain motivational conditions should be systematically related to the selection of specific decision options (e.g., the nuclear option). Thus the only additional assumption about decision making required to pursue this particular line of inquiry is that, by whatever process stimuli and responses are linked, the relation between them is probabilistically consistent over both time and space.

In sum, the motivational hypothesis posits that *decisions to initiate nuclear weapons programs should be systematically related to prevailing motivational conditions.*

NUCLEAR PROLIFERATION IN A SUI GENERIS WORLD

Many people in both the academic and the policy communities argue persuasively that national and international political phenomena— such as coups and revolutions, wars, and decisions to acquire nuclear weaponry—are one-time events. While analogies and similarities certainly exist, nothing is ever really the same when you take a close look. In a sui generis world, the pattern is that there is no pattern. Events take place (or fail to take place) because myriad factors coalesce for an instant that will never be repeated.

Here the best one can hope for is to understand why past proliferation decisions were made by examining each in detail. While technology is seen as a necessary condition for future proliferation decisions, the remaining necessary conditions are neither identifiable, nor predictable a priori, nor consistent. Proliferation decisions that occur will do so because specific governments perceived themselves to be in specific contexts and because the individuals and institutions involved ultimately favored the "nuclear option."

The pivotal role of the idiosyncratic in the sui generis hypothesis does not necessarily imply rejection of the notion that certain underlying conditions are necessary for events of a given type to occur. Rather, the primary thrust of the hypothesis has to do with why events occur precisely when they do. Here it might be useful to think in terms of necessary conditions and triggers. While the assassination of Archduke Francis Ferdinand was the trigger for the First World War, it was hardly the cause. Had certain prevailing conditions not existed, the killing most certainly would not have engulfed the world in flames. On the other hand, the necessary conditions for a world war had existed for a considerable time before the outbreak of hostilities. It failed to occur sooner because no triggering event had been forthcoming. In a similar fashion, decisions to "go nuclear" may be the product of prevailing conditions and largely irrelevant triggers. Thus no two decisions can be treated as a generic class, and hopes for being able to generalize from history, or forecast the near future, will remain unfulfilled.

Precisely because of its distinctly negative perspective, this hypothesis will serve well as the *null hypothesis*. Though it can, and will, be tested in an explicit fashion, failure of the several variations of the technological imperative hypothesis and the motivational hypothesis will be strong evidence for accepting this hypothesis of a sui generis world.

All three classes of hypotheses described above recognize that it is human decision makers who choose whether to manufacture nuclear weapons. Atomic bombs don't build themselves. All three acknowledge that political decisions (and therefore motivations) and technology have roles to play. Where they differ is in their attribution of primary influence: what is necessary versus what is sufficient for a decision to proceed with nuclear weapons production. Of course anecdotal and impressionistic "evidence" can be mustered in any number of ways to support or to refute these three general hypotheses. Flinging and dodging anecdotes is, however, a poor approach to analysis. Proper investigation of these hypotheses must incorporate *all* relevant cases into the data base—not just those that seem to support a particular line of argument. Again, the goal is not to "prove" which hypothesis is right or which is wrong. There is probably a touch of truth in all three classes of hypotheses. Since all three types incorporate the same variables—but weigh them differently—individual examinations of each hypothesis at least should shed considerable light on interactions among the variables and how they behave. It would be worth knowing, however, whether any individual model of the nuclear proliferation process dominates the others. Which might best account for most of what we have observed in the past three decades or so of the nuclear era? Since only limited information is available, what will tell us the most about past and future?

2

The Technological Basis of Nuclear Proliferation

No aspect of nuclear proliferation has been dealt with so comprehensively as the technology of nuclear weapons and nuclear weapons production. The intricacies of designing and building plutonium production reactors and reprocessing plants and the problems and pitfalls of enriching uranium have all been discussed in great detail. Options for diverting nuclear fuel from civil nuclear power facilities have received the most attention of all. And whatever can be said in the open literature (and probably some things that should not) about designing and building an atomic bomb has already been said.[1]

What has not received systematic treatment is a method for using intelligence on technical capabilities to determine whether a given country is in fact capable of nuclear weapons production. What is required technically and industrially for a country to successfully initiate and carry out a nuclear weapons program? Any attempt to examine why a nation "goes nuclear" must begin by determining when the option first became available. At what point did a country first cross the threshold of acquiring a latent capacity?

To answer these questions, we need an explicit definition of "a nuclear weapons program." Definitions and usage of the terms "nuclear weapons program" and "nuclear capability" vary considerably within the literature. For some analysts a "nuclear weapons program" necessarily encompasses the development of a host of related, though independent, capabilities such as nuclear warheads, advanced delivery vehicles, secure command and control facilities, an operative "strategic doctrine," assured second strike systems, and so forth. Often the massive American effort is cited as the stereotypical nuclear weapons program. Others have posited prospective nth country programs of less ambitious dimensions—the

19

simplest being characterized by the singular ability to put together a marginally "fissile" (dud) explosive device. Just as the Third World conventional military forces need not be of the dimensions of the American or Soviet armed forces to be regionally significant, so may Third World nuclear forces set their own standard.

Correspondingly, the differences between so-called optimistic (slow growth) proliferation forecasts and pessimistic (rapid growth) proliferation forecasts are to a large extent influenced by what they posit a nuclear weapons program to be. In other words, how one defines a nuclear weapons program will in many ways determine which nations appear capable of becoming nuclear weapons countries at any particular time. This in turn will affect one's perception of the urgency of the "proliferation" problem. Efforts to integrate prior research into a robust and cumulative body of knowledge have been hampered by inconsistencies in the specification of the supposedly common subject of investigation—a nuclear weapons program. The nonproliferation policy debate has suffered equally, if not more. Precisely because this definitional issue has produced so much confusion, it is important to digress for a moment and try to reach some agreement on what a "base case" nuclear weapons program might look like. In determining this common denominator, answers to the following questions would prove useful: Is there some minimum number of nuclear weapons that must "roll off the production line"? Must a nuclear weapons program be planned to provide for a continuous production schedule, or could it be terminated? Is the development of an advanced delivery system an integral and necessary part of such a program? Is the testing of prototype nuclear explosive devices necessary? And should a "peaceful nuclear explosives" (PNE) program be considered equivalent to a nuclear weapons program?

How Much Is Enough?

Historically, the size and scope of past nuclear weapons programs have varied substantially. The initial American program had a potential production rate exceeding thirty fifteen-kiloton weapons per year, including both plutonium-based and uranium-based weapons. Moreover, no fewer than five distinct fissile material production techniques were simultaneously employed. In contrast, the intitial British program could muster only some eight to ten fifteen-kiloton weapons per year, all based on plutonium. The initial French program appears to be in many ways a carbon copy of the British approach. The initial Soviet nuclear weapons production rate fell somewhere close to that of the British and French. The initial Chinese production rate was probably also comparable to that of the British and French. It is tempting, therefore, to inquire whether historical experience might provide guidance on individual countries' requirements for the size and scope of their nuclear weapons program. Is there some underlying logic that a priori determines how much is enough?

With regard to America's Manhattan Project, the official record clearly attests that its eventual size and scope were the products of chance—not planned goal seeking. During the course of the Manhattan Project, two independent lines of nuclear weapons were designed—plutonium-based and uranium-based. This "dual birth" evolved not from an interest in technological diversity but rather from an inability to decide which approach had the greater likelihood of working. Moreover, within each of these design approaches a variety of competing—but, more important, redundant—fissile material production techniques were pursued as hedges against "dead ends." As a result, in the course of producing U-235-based weapons, no fewer than four distinct methods of isotope separation were developed, tested, and engineered—three were actually employed. Both of the weapons design approaches and all the individual production techniques proved successful. Thus the scope of the American program was simply a function of the great uncertainty regarding how to proceed and America's corresponding ability to pursue every conceivable avenue of research and production.

Likewise, the initial production rate (size) of the United States program was mechanistically determined by this overlapping multiplicity of production facilities that (unexpectedly) proved successful. In a most revealing narrative, Leslie Groves (the former United States Army general who directed the Manhattan Project) recalls that during a meeting with project scientists:

> I asked the question that is always uppermost in the mind of an engineer: with respect to the amount of fissile material needed for each bomb, how accurate did they [the scientists] think their estimate was? I expected a reply of "within 20 or 50 percent," and would not have been greatly surprised at an even greater percentage, but I was horrified when they quite blandly replied that they thought it was correct within a factor of ten.
>
> This meant, for example, that if they estimated that we would need one hundred pounds of plutonium for a bomb, the correct amount could be anywhere from ten to one-thousand pounds. *Most important of all, it completely destroyed*

any thought of reasonable planning for the production plants for fissile materials. . . . This uncertainty surrounding the amount of fissile material needed for a bomb plagued us continuously until shortly before the explosion of the Alamagordo test bomb on July 16, 1945 [emphasis added]. (Groves 1962, 40; see also Brown and McDonald 1977).

As it turned out, the amount of fissile plutonium or uranium required was considerably less: roughly ten pounds of Pu-239 or thirty-five pounds of U-235.

In sum, there is strong evidence that besides the goal of producing *a* workable bomb, the scientific, technical, and production uncertainties that surrounded the early American project left little room for a priori nuclear weapons stockpile planning.

In the British case (1947-52),[2] however, there was an apparent effort to develop a rational calculus for determining an optimum production rate. First, consideration was given to project size of the (suspected) secret Soviet nuclear weapons program. How could Britain expect to remain a global power in the postwar era unless it established and maintained a comparable nuclear weapons production rate? Second, in contemplating a battlefield nuclear defense strategy for Western Europe, the British Joint Chiefs attempted to estimate commensurate nuclear explosives requirements. (In fact, they simply "scaled up" previous estimates of the number of such weapons that would nominally be required to destroy Great Britain!) This determined the lower limit of nuclear weapons required as a function of purely military criteria. Finally, consideration was given to likely United States nuclear weapons production rates and the corresponding proportion of America's nuclear weapons that could be expected to be allocated to European defense.[3] That is to say, the planners hoped the Americans would eventually pick up part of the nuclear burden of defending Western Europe.

The resulting calculations suggested that Britain needed approximately six hundred nuclear weapons. Positing that the United States would furnish about two-thirds of these, the British estimated that they would have to produce roughly two hundred weapons over five years (1952-57)—forty bombs a year.

But the British also maintained high hopes for potential industrial applications of nuclear energy (Gowing 1974, chaps. 6-12). Given the deteriorating state of their national economy, the British were forced to choose between constructing enough plutonium production piles to produce the forty bombs a year or building a low-separation uranium enrichment facility (to fuel nuclear power reactors) at the expense of some bomb-production capacity. They chose the latter course,[4] and Britain's actual rate of nuclear weapons production hovered between eight and ten weapons a year.[5] Thus, in the final analysis it seems that the rational approach to military force planning implied by the strategic analysis effort was simply over-shadowed by other economic-industrial considerations and the constraints of limited resources.

In the French case (1956-60), one is particularly hard pressed to find any evidence of a strategic-military rationale underlying the initial production size or scope (Goldschmidt 1962; Kelly 1960; Kohl 1971; Scheinman 1965; Zoppo 1964). However, it is interesting that the aggregate plutonium output from the three French plutonium production reactors (G1, G2, G3 reactors at Marcoule) was equivalent to that of Britain at the time the French plutonium production reactors were being designed. Perhaps this is a coincidence, perhaps it reflects a general technical and resource limit on new nuclear weapons, or perhaps the French simply wished to demonstrate a technological capacity equal to Britain's.

Not surprisingly, no information is publicly available on what criteria the Soviets or the Chinese (People's Republic of China) used to determine nuclear weapons production rates or program scope.

If any inferences regarding a hypothetical "base case" nuclear weapons program can be drawn from the historical record, then, they are indeed minimal. The case of the United States program is obscured and confounded by its innovative and experimental aspects. In the British case there is some evidence that negative constraints were the most potent determinants of production rates and program scope. For the French nuclear weapons program there are several hunches but no strong evidence.

As opposed to trying to make a point prediction of program size for each nation, we could take a somewhat more abstract approach and posit an interval—a range of points—bounded by a lower threshold. In this way we could draw upon more theoretical political and military notions about the potential utility of nuclear weapons. Specifically, it might be possible to estimate "how much is definitely *not* enough." I have already pointed out, for example, that the British settled for a nuclear weapons production rate less than 25 percent of what they deemed desirable. Here too the United States was prepared to "accept" a production rate of only several nuclear weapons a year from its own program. But how much divergence from that target production rate would they have been willing to accept and still pursue program development?

The notion of a minimum threshold for the size of a nuclear weapons program is predicated on two interlocking assumptions: (1)

given the desire to "go nuclear," any prospective nth country would either develop a program of at least some minimum size, or (2) if that mimimum size could not be attained, no program would be initiated. This is essentially a "minimum utility" concept. Theoretically this formulation is fine. But a question remains whether it is possible to derive a specific value that will more or less correspond to average international perceptions of a minimal utility threshold for the size of a nuclear weapons program.

In this respect, before the new safeguards guidelines under the Non-Proliferation Treaty, the International Atomic Energy Agency (IAEA) exempted from inspection any reactors of less than three megawatts-thermal (MWth) output—up to two such reactors per country (see INFCIRC/66/Rev. 2, in Sanders 1975, 98). Employing similar criteria, the United States Congress enacted PL 93-485 (October 1974), reserving the right of full congressional review of any United States reactor sales with a thermal output greater than five megawatts (Barber Associates 1975, IV-53). Both these "regulations" imply that the authorities did not consider the proliferation-related threat of reactors of five to six megawatts substantial. Such reactors would theoretically be capable of producing enough plutonium to make one fifteen-kiloton bomb every four or five years. This suggests that the base case program size be set at 0.25 (or 0.2) bombs a year. Over the course of a decade, then, an nth country following such a program would expect to accumulate two to three fifteen-kiloton atomic weapons. Such a production rate might satisfy, for instance, "weapon of last resort" criteria for nations like Israel or Taiwan. It might also be fairly easy to conceal—thus enhancing the "surprise" aspect of the weapon's military utility.

However, a number of previous studies of nuclear proliferation have tended to settle subjectively on a minimum production threshold somewhat higher—roughly one fifteen-kiloton weapon a year (Barber Associates 1975; Beaton and Maddox 1962; United Nations 1968; Office of Technology Assessment 1977). Still others argue that even within a regional setting one bomb a year would not be politically or militarily significant. For these analysts, something closer to five to ten bombs a year seems appropriate.

The extreme fuzziness of this issue implies that any attempt to make a conclusive judgment would be in part an exercise in inferential heroics. However, a good case can be made for initially considering a production rate of one bomb a year. The preference for one bomb a year, rather than 0.25, is based on the fact that the relative resource demands of the former are not significantly greater than those of the latter, quantitatively or qualitatively. For a disproportionately small increase in the fraction of the national resource base "diverted" to these demands, a prospective nth country could quadruple its production output rate from 0.25 to 1.0 nuclear weapon per year.

The choice between one bomb and, say, seven bombs a year is based more on symbolics and psychology than on economic or military analysis. Many students of proliferation have come to the conclusion that:

> the danger of proliferation today is the emergence of what might be called the primitive nuclear powers with a *limited* stock of untested nuclear weapons [emphasis added]. (Olgaard 1969, 219)

> One must be careful to consider thoughts about physical realities, as well as the realities themselves. Even if physical developments pose no *real* problem, the fears they bring about can take on a life of their own [emphasis added]. (Quester 1973, 210)

So, in this sense:

> even a minimum size, inefficient and unreliable "bomb" could result in a local disaster and an international crisis of immense proportions. (Barber Associates 1975, V-5)

A one bomb a year program also seems to drive those studies of nuclear proliferation that concentrate on the "diversion option"— that is, a country "goes nuclear" by diverting spent reactor fuel from a power reactor, reprocessing it, and using the plutonium to manufacture an atomic bomb. In this context the range of assay (safeguards) techniques generally available can ensure accountability within a few percentage points of total inventory. In other words, if a typical power reactor produces almost three hundred kilograms of plutonium a year (Dawson, Deonigi, and Eschbach 1965, 101-5; Willrich and Taylor 1974, chap. 3) and assay inaccuracies are about 2 percent (Leachman and Althoff 1972), this suggests that about six kilograms of plutonium could conceivably be diverted without triggering a safeguards alarm—almost enough to make a single fifteen-kiloton bomb. Thus, by setting the minimum program size at one weapon a year, this study is sensitive to the arguments of both those who have emphasized the civil nuclear power diversion option and those who have emphasized dedicated programs in a Third World political, military, and industrial setting.

Chapter Two

CONTINUOUS VERSUS TERMINATE PRODUCTION SCHEDULES

The question now arises whether the national commitment to the production of nuclear weapons need be unceasing. Must the program be capable of producing weapons indefinitely into the future? This line of inquiry derives its importance from the issue of resource inputs; if a nuclear weapons program must "go on forever" then crucial resources must remain available.

In this respect, the United States-Soviet strategic nuclear arms race, and the concurrent process of vertical nuclear proliferation among the United Kingdom, France, and the People's Republic of China, seems to support the proposition that a nuclear weapons program is a never-ending production line. In essence, once is not enough. It is conceivable, however, that, from the vantage point of future nth countries otherwise unencumbered by the strategic-military influences of the American-Soviet nuclear rivalry, an endless output of nuclear weapons might be regarded as unnecessary—if not unmanageable. This is not to say that the nuclear weapons production capacity would be dismantled. Rather, the program might be suspended and placed in standby readiness until some future condition is met. Such a condition might include, for example, a perceived change in the quantity of nuclear weapons needed for national defense or, in a somewhat different vein, an increase in the availability of some crucial resource.

I raise this possibility not in the belief that there is some magic level of nuclear "satisfaction," but rather to sensitize the reader to the notion that a prospective nth country might initiate a nuclear weapons program with the clear knowledge that the finite availability of specific resources could constrain the size and scope of the intended production. In particular, the perceived need for a few operational atomic bombs produced today in a clandestine program might outweigh the threat of international sanctions should the program be discovered. This has certainly been the line of argument employed by the Israelis in their indictment of Iraqi nuclear ambitions. The Iraqis, they argue, would willingly accept international sanctions and the shutdown of their research reactor facility if they could first use the enriched uranium fuel to make several nuclear weapons. Likewise, Libya's Qaddafi is reported to have gone to great lengths to purchase just a single nuclear weapon. Thus there is anecdotal evidence that a finite stock of nuclear weapons may be perceived as having sufficient utility to permit a proliferation decision.

TESTING

Within the context of a nuclear weapons program, the issue of "testing" stems from both technical and decision-making considerations. First, consider the technical aspect of testing. If an operational test of prototype devices is indeed required to attain some minimal level of technical and political confidence in the development and manufacture of nuclear weapons, then testing itself becomes an inherent part of the program. A logical extension of this argument, moreover, implies that an inability to conduct such tests translates into an inability to undertake a nuclear weapons program (Hohenemser 1962, 269; Beaton 1966b, 38-39). To be sure, "Not all nations are provided with deserts within their boundaries. Uninhabited oceanic islands are not always accessible. . . . Where would a Swiss bomb, or a German bomb, be tested? In these and other cases *the need* to actually test atomic weapons would be a serious obstacle in the attainment of nuclear power" (Beaton and Maddox 1962, 20; emphasis added).

The notion that testing is an essential *technical* element of a nuclear weapons program was one of the basic arguments for a Comprehensive Test Ban in the late 1950s and early 1960s (Jacobson and Stein 1966). And, more recently, the international diplomatic offensive that purportedly halted an imminent South African nuclear test in 1977 was equated with preventing South Africa from embarking on an atomic weapons program.[6]

But the question remains: Is there a *technical need* for prospective nth countries to test protonuclear bombs? Does an inability or unwillingness to test a nuclear explosive device mean that a nation cannot pursue nuclear weapons production? The current informed consensus seems to be that, given that some nth country adopts one of the tried-and-true designs, there would be few compelling reasons not to expect its first bomb to work (Kramish 1962, 23; Wentz 1968, 23-24; Quester 1973, 39; Brennan 1976, 108). The United States did, after all, employ an untested uranium bomb against Japan. And as far as is publicly known, the first atomic weapons tests of all five recognized nuclear powers, as well as India's first PNE test, were successful. Furthermore, from a scientific and engineering standpoint, fabricating the weapon is one of the lesser challenges of the entire enterprise. The necessary "physics" is easily accessible in the open literature, and the requisite conventional explosives techniques have long been employed in industrial processes (Olgaard 1969, 219-20; United Nations 1968, 58; McPhee 1974). The most critical components (e.g., the high-explosive detonation system) can all be tested without the fissile material. It is hard to imagine, then, how a dedicated team of competent nuclear physicists and engineers could not devise a moderate-confidence nuclear weapon design.[7]

Clearly, testing *would* allow for the "tightening up" of design specifications, permit greater efficiencies in terms of energy yield, and offer an opportunity to experiment with alternative designs. But this borders on a technical luxury, not a technical necessity.

The issue of testing has also been raised on the basis of politico-military considerations. It has been argued that nuclear weapons provide military security through deterrence. Deterrence, however, demands communication of "the threat" to the target party—and testing achieves this. Thus without a test, or series of tests, to establish the existence of a national nuclear weapons arsenal, the deterrent potential may be negated. I say "may be negated" because a deterrent effect can be obtained in other ways. Israel, for example, has gotten "nuclear" mileage from its ambiguous status—propped up by little more than rumor.

There also are other military applications for nuclear weapons. Specifically, the battlefield nuclear role and the "weapon of last resort" role are two options that could have particular appeal to small military powers. In such circumstances testing would "tip their hand," possibly compelling their antagonists to "go nuclear." Alternatively, a secret nuclear stockpile would maximize the element of surprise and thereby confer maximum military utility,[8] which testing might decrease.

From a broad perspective, testing is not a technical necessity for an operational nuclear weapons program. While testing may have many useful aspects, it should not be considered an integral part of that program.

ADVANCED DELIVERY SYSTEMS

The question whether a nuclear weapons program must develop advanced delivery systems has also received considerable attention. In particular, throughout the 1950s and early 1960s there was extensive discussion on the relative significance of a nuclear weapons stockpile lacking an advanced means of delivery. Without a commensurate fleet of strategic bombers or ballistic missiles, it was argued, the military and political value of a nuclear weapons arsenal would be dubious at best.

But one need only recall the United States reaction to the first Soviet atomic test in 1949 to see the weakness of this argument. Here was a country halfway around the world that literally possessed no meaningful way of delivering its newly acquired nuclear weapons. Yet its first test provoked a major reassessment of the United States military posture, and United States capabilities to produce nuclear weapons in particular. It was assumed that the Soviets would quickly attempt to procure the requisite delivery vehicles.

Most previous studies also took it for granted that *the targets* of nth country nuclear forces would include either the superpowers themselves or the superpowers' well-armed allies. As such, considerations of antiaircraft and antimissile systems naturally arose. A more broadminded perspective, however, notes that "the principal sufferers from nuclear spread would be the countries and the people in those areas in which the infection takes place" (Schlesinger 1967, 11-12). Here we are talking not about symmetrical or asymmetrical nuclear confrontations involving at least one militarily sophisticated major power but, on the contrary, about conflict between less well-endowed countries. Some have called these nations nuclear "pygmies" (Betts 1977). In a regional context, even a comparatively primitive nuclear force of several fission weapons and several jet aircraft may accrue considerable military and political significance. In an emergency situation, it is conceivable that passenger jets or cargo transport planes might be used as "delivery vehicles." Indeed, Israel has long maintained its commercial aircraft in a dual-status (military) role and used them as transports during the 1973 Middle East War. The United States military has similar plans to use America's commercial airline fleet in an emergency airlift role to Western Europe. Clearly, an aircraft that can transport tanks, ammunition, and men into a combat area can likewise carry nuclear weapons. The Boeing 707 is a more sophisticated aircraft than were America's B-29s.[9] Moreover, nuclear weapons buried along key invasion routes—acting as nuclear mines and barriers—need not be delivered at all.

One must also take into account the tremendous growth in the international conventional arms trade. Many "small" countries today can purchase—and have purchased—advanced antiaircraft systems. A "jetliner bomber squadron" therefore could be a high-risk method for delivering a precious cargo of atomic bombs. However, the international arms trade has not been confined to defensive weapons systems. In particular, one is hard pressed to find any country in the position to contemplate nuclear weapons manufacture within the next fifteen to twenty years that is without a *regionally* effective delivery capability (International Institute for Strategic Studies 1977). In the end, "technologically backward nations have the advantage of a much cheaper, unsophisticated defense environment" (Kaul 1974, 7).

There may be, moreover, less orthodox ways to deliver a nuclear weapon to its target, particularly in a regional setting: "it is a myth that for any country to attack any other country requires anything like the sophisticated long-range missiles that we have. There have been rather detailed (classified) studies . . . of which the overall results . . . would lead one to a conviction that any country in the world could attack any other country by means other than long-range bombers or missiles, using clandestine infiltration operations of various kinds."[10] Delivery by land vehicle or by sea (to coastal targets) should not be ruled out. In essence, the framework established by the United States-Soviet intercontinental nuclear rivalry—a rivalry characterized by long-range bombers and ballistic missiles—may well be the exception rather than the rule and should not be allowed to bias perceptions of other potential international nuclear threat relationships.[11]

It seems reasonable, then, to posit that, though an advanced delivery capability may become critical as a nuclear weapons program moves into its advanced stages, during the initial phase "how to get the bomb from here to there" is overshadowed by "how to get the bomb" (Hohenemser 1962; U. S. Senate 1968, 30-31). In sum, we can assume that by the time a prospective nth country acquires a latent capacity to manufacture nuclear weapons a rudimentary delivery capability can be taken for granted and therefore should not be considered an integral part of the nuclear weapons program.

"NONWEAPON" NUCLEAR EXPLOSIVES

There is a class of nuclear explosives—so-called peaceful nuclear explosives—that are purportedly nonmilitary. Their alleged purpose is for nonmilitary engineering—a "more bang for the buck" approach to excavation and mining.

Assertions by nations like India and Brazil aside, the plain technical fact is that, in terms of their capabilities and effects, PNEs and nuclear weapons are indistinguishable. A former director of the United States Arms Control and Disarmament Agency, Fred C. Ikle, has observed: "We do not now have a way of discriminating between tests of a nuclear device of a country that is just beginning a test program, discriminating whether it is for peaceful purposes or military purposes. A country that is just beginning a program of testing nuclear explosives will inevitably learn something about nuclear explosives, and such an explosive is a very destructive device that could be used for military purposes" (U.S. House of Representatives 1975, 225-26).

One need only note the explicit relation between the Threshold Test Ban Treaty and the PNE Treaty to confirm this point.[12] Moreover, photographs of the Indian PNE reveal that it could easily have been transported to a military target. There can be little doubt that, from a military standpoint, a PNE detonated in proximity to any target—whether a city or an armored column—would produce devastation identical to that of an appropriately labeled atomic bomb of equal yield. In other words, the ability to manufacture PNEs is *indistinguishable* from an operational capability to produce nuclear weapons. Thus exploratory PNE programs can be viewed as surrogates for an initial nuclear weapons program and a latent capacity to manufacture PNEs is also a latent capacity to manufacture nuclear weapons.

ASSESSING LATENT CAPACITIES

To summarize the discussion above, for present purposes a nuclear weapons program is any nationally directed program with the objective of producing at least one functional nuclear explosive device a year, averaged over several years. Whether or not simultaneous planning for delivery vehicles takes place or testing of any nuclear explosive devices is scheduled is taken to be external to the nuclear weapons program itself. Accordingly, a nation is said to possess a latent capacity when there is reasonable confidence that such a program is feasible.

What does feasibility entail? "Going nuclear" involves diverting some portion of the national resource base (capacity) to satisfy the demands of the activities associated with the nuclear weapons program. Whether some particular country, at a specific point, is or is not capable of producing nuclear weapons will therefore depend on the extent to which it can satisfy every one of the program-specific resource demands. What is in fact being questioned is *relative feasibility*—the ability of a specific country to meet a range of resource demands (dictated by the needs of the program activities) with a finite resource base. The notion of relativity is invoked when we recognize that national resource capacities vary with time and space and that the resource demands of a nuclear weapons program itself may vary with time.

The term "resources" should be interpreted in its broadest sense—as not only material resources (uranium, steel, concrete, water, etc.), but scientific and technological (engineering) expertise and manpower, industrial production capacity, and capital availability as well. Unless the specific resource demands from each of these areas can be satisfied, producing nuclear weaponry will be impossible.

To this end, let me digress for a moment to discuss these four general classes of resources in greater detail. In particular, there has been a tendency to blur the distinction between the scientific demands of nuclear weaponry and the technological demands.

The scientific demands of nuclear weapons production are characteristically both qualitative and quantitative. They are qualitative in that they entail some basic level of knowledge. They are quantitative in that some minimum (critical) number of individuals must be acquainted with that knowledge (Office of Technology Assessment 1977, 174-81). Specifically, the science that underlies nuclear weaponry is a diverse body of knowledge encompassing the fundamental principles and processes that make the release of nuclear energy possible. This subsumes not only nuclear physics and nuclear chemistry but classical mechanics, thermodynamics, kinetic theory, and the chemistry and metallurgy of transuranic elements as well. A national science base lacking in any of these areas could not support a nuclear weapons program.

Technology, in general, can be thought of as "the *application of science* to the manufacture of products and services. It is *the specific know-how required . . . to design the product, and to manufacture it.* The product is the end result of this technology, but it is not the technology" (Bucy 1977, 28-29; emphasis added). While the science of nuclear weaponry defines the physical and chemical principles and processes through which the requisite materials for an atomic bomb can be produced, and under which the bomb itself can be expected to work, the technology of nuclear weaponry encompasses the specific engineering "know-how" required to design and construct the various production facilities and ultimately the components of the device.

This distinction between scientific demands and technological demands is crucial to any analysis of the relative feasibility of nuclear weapons production. The scientific aspect reveals the particular conditions needed for producing an atomic bomb, thus specifying what needs to be done. The technological aspect then specifies how those particular conditions are met—the actual mechanics. Thus something can be scientifically feasible in that there may be a thorough understanding of what conditions need to be created, yet technologically infeasible in that existing "know-how" is not sufficient to create those conditions.[13] Controlled nuclear fusion for electrical energy production is an example of an application where scientific feasibility exists but technological feasibility remains to be achieved in the distant future.

The resource demands related to industrial production capacity and material resources are almost self-explanatory. A prospective nth country must possess the ability to manufacture many of the relevant input products and construct the necessary plant facilities.[14] And, perhaps more important, a basic industrial infrastructure would be necessary to maintain and service the plant facilities. There are, of course, both quantitative aspects (requisite input and output levels) and qualitative aspects (product quality) associated with the industrial production demands. Quantitatively, the national industrial capacity would have to be large enough to satisfy the program's demands while fulfilling other national needs. For example, the American Manhattan Project consumed almost as much electricity (between 1944 and 1945) as France's entire electrical generating capacity during that period. A prospective nth country's industry must also be able to provide the proper quality of products. So-called nuclear-grade graphite, for instance, is more difficult to manufacture than commercial forms of graphite.

Then, too, access to material resources—raw materials and semifinished products (and, in the case of an incomplete national production capacity, finished materials like stainless steel)—of sufficient quantity and quality are obviously essential to keeping the program operating. The availability of capital resources to finance the program must also be considered. Thus, assessing latent capacities requires that all these factors be taken into consideration.

Of course some resource demands can be satisfied through imports. The oil-rich Arab states, for example, have imported large quantities of raw materials, semifinished and finished products, laborers, technicians, scientists, and engineers in support of domestic industrial development. Yet imported material and labor can be used only to plug certain holes in the national resource base. Often ignored in surveys of the technical requirements for nuclear weapons production is what might be called the "Sears, Roebuck" factor—the ability to rapidly replace worn, broken, missing, and otherwise unusable items. Many technology transfer projects to developing countries have floundered because of continual long delays in replacing components. Many analysts take for granted ready access to sufficient stocks of standard industrial items: steel piping, concrete, gauges, electrical cable, plumbing fittings, and so on. Indeed, by placing undue emphasis on the top of the capability pyramid (the "special" technical aspects) while ignoring the bottom, many have overestimated the nuclear potential of states such as Pakistan and Egypt.

Along with the "Sears, Roebuck" factor we should add the "Yellow Pages" factor—the ability to rapidly locate plumbers, electricians, masons, welders, machinists, and such, for maintenance and repair. While skilled craftsmen can be imported for construction, a substantial reservoir must also be kept available for routine service. This is where many prospective proliferants fall short. They may be able to "buy" the limited numbers of engineers and scientists and "special" technical components required for a nuclear weapons program, but they cannot meet the more mundane labor and material demands. One must also consider management aspects. As with any large industrial project, coordinating a nuclear weapons program requires substantial managerial and administrative talent. Developing countries often hire foreign engineering firms to manage and administer large-scale industrial development. It is doubtful, however, that Saudi Arabia would hire the Bechtel Group to oversee the Saudi nuclear weapons program. In sum, regardless of the prospects for importing critical human and material resources, considerable attention must still be given to the underlying industrial resource base of the countries being examined.

With these considerations in mind, the question remains: How can we determine if and when countries acquire latent capacities? Two alternative strategies present themselves. On the one hand, one could collect subjective assessments by specialists and experts. Methods like the "delphi technique" are available for codifying the impressions, insights, and opinions of those most familiar with individual countries—in particular, impressions of technological capabilities. The problem with subjective assessments is that they are very sensitive to the idiosyncratic biases, prejudices, and preconceptions of the respondents. As I have noted, there is considerable confusion and disagreement regarding what a nuclear weapons program is (or is not), and correspondingly what is required to manufacture nuclear weapons. And all too often one finds highly divergent views of a country's level of technological development. One also finds many instances where assessments of a country's intent to produce nuclear weapons affect evaluations of its capabilities to do so. There is a tendency to attribute greater capabilities to countries perceived as interested in nuclear weapons. This bias is particularly pronounced for less technologically developed countries. Unfortunately, the corrective filtering that methods like the delphi technique are supposed to provide is not likely to work in these instances because individual experts will be making assessments of different countries.

Alternatively, a list of observable technical indicators might be devised that could systematically discriminate between nations capable and incapable of manufacturing nuclear weaponry. One would use indicators similar to those that go into expert subjective assessments. The crucial difference is that, by indexing the list of indicators explicitly, one ensures that the same criteria are systematically applied in all assessments. This approach may well overestimate a specific country's capabilities. But systematic biases are fairly easy to detect and correct—at least compared with the idiosyncratic errors likely to be found in subjective assessments.

In fact, when both strategies are used together the systematic biases of one offset the idiosyncratic biases of the other. The result should be smaller overall errors than when either is used alone. First a base case nuclear weapons program is outlined—a bare-bones undertaking that employs "low" technology just sufficient to enable a country to manufacture nuclear weapons. Assuming a nation has no nuclear infrastructure—no preexisting facilities for nuclear research or nuclear power production—what is the least difficult and least demanding (or most feasible) path to producing one crude but functional nuclear weapon a year averaged over several years? The corresponding resource demands of the base case nuclear weapons program could then be derived and used for guidance in selecting technical indicators of national capabilities to meet those resource demands. Subjective assessments and historical case analyses can be used both for guidance on indicator construction and for validation. These indicators subsequently can be used to estimate the earliest date at which a given nation, starting from scratch, could satisfy the technical resource demands of the base case nuclear weapons program.

Once this has been done, the implications of various countries' nuclear infrastructures could be evaluated and factored into the analysis. Nations that possess significant nuclear infrastructures (e.g., plutonium production reactors) before embarking on a nuclear weapons program would already have satisfied certain of the resource demands of the program. In such instances determination of latent capacities would simply exclude from consideration those technical and economic resource demands components satisfied by a nation's existing nuclear-related infrastructure. These "adjustments," in turn, reduce the set of indicators that need to be examined.

Base Case Nuclear Weapons Program

Since the resource demands are driven by the plants associated with fissile material production and processing, the discussion that follows focuses on fissile material facilities. For a nation with latent capacity but lacking a preexisting nuclear infrastructure, there is one most direct path toward developing an indigenous atomic weapons production capability. This method employs an air (gas)-cooled, graphite-moderated, natural uranium fuel production reactor (GCR) in conjunction with a laboratory scale plutonium reprocessing facility to yield enough Pu-239 for one fifteen-kiloton weapon per year. While there are other routes to plutonium production (e.g., a heavy water moderated production reactor) or uranium enrichment for producing bomb-grade uranium, any nation capable of following these other paths would certainly be capable of carrying out this base case nuclear weapons program. Thus it really does represent the "common denominator" of initial nuclear weapons programs.

Figure 3 diagrams what the base case nuclear weapons program

Fig. 3. A parameterized flow chart of an atomic weapons program. For a detailed discussion, see Appendix B.

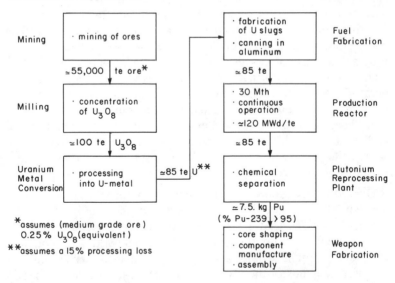

entails. The key to the production program is the thirty-megawatt-thermal production reactor. The mining, milling, chemical conversion, and fuel fabrication stages represent "support" activities for the reactor. Likewise, the plutonium reprocessing stage can be viewed as a "postirradiation" support activity. Thus the various parameters associated with each of these "support" activities are determined by the parameters associated with the production reactor. For one fifteen-kiloton plutonium bomb a year of primitive design, it can be assumed that roughly seven kilograms of plutonium (over 90 percent Pu-239) would be required (Beaton 1966a, 38; Brown and McDonald 1977, 338; Van Cleave 1974, 41). Accordingly, a 30 MWth GCR, operated at 80 percent capacity over the continuous period of one year, and with a conversion ratio of 0.8 (Pu/U-235 fission), would produce roughly 7.5 kilograms of plutonium a year. With a required fuel burnup figure between 110 Mwd/te and 120 Mwd/te this would ensure a high (over 95 percent) relative fraction of Pu-239. Following this base case program, a country without any nuclear-related infrastructure will require, on average, roughly six years from program start to produce the first bomb (Wohlstetter et al. 1979, 42-46). So our operational definition of a (no-infrastructure) latent capacity is: the possession of a national resource base that, mobilized for a nuclear weapons program, would produce its first nuclear weapons within six years of initiating efforts.

Countries with a *moderate infrastructure* might be able to satisfy roughly one-third to one-half of the resource demands of the base case nuclear weapons program using preexisting facilities. Three specific forms of moderate infrastructures are likely to be encountered. First, existing research reactors could be used to produce nuclear weapons-usable plutonium. India's CIRUS reactor, for example, is believed to have been the source of plutonium for its PNE core. (Although the Iraqi reactor destroyed in the Israeli raid of June 1981 could have been used to breed plutonium, its highly enriched uranium fuel could have also been used to produce several nuclear weapons.) A second approach could be the diversion of power reactor fuel. Moderate-size power reactors can produce ten to twenty times as much plutonium annually as the largest research reactors. The third form of moderate nuclear infrastructure is the less likely situation of a country with a large hot cell or pilot-scale reprocessing facility but without reactors of substantial plutonium-producing value. Both Argentina and Belgium were in this last category sometime in the past. The corresponding "time to bomb" would most likely be cut from the six years of the base case program to four years (Wohlstetter et al. 1979, 42-46).

Countries with *advanced nuclear infrastructures* might possess everything they need to initiate the manufacture of nuclear weapons—except the weapons fabrication plant and the nonfissile weapon components. Here there are basically two alternative forms of advanced nuclear infrastructures: either a country has both a plutonium producing reactor and a reprocessing facility (or large hot cell), or it has a uranium enrichment plant. In either case, almost all the nuclear weapons program resource demands would be satisfied a priori. The Canadians, for example, have had plutonium-producing reactors and a reprocessing facility since 1947. If, say, in 1980 a decision had been made to launch a Canadian nuclear weapons program, clearly none of the activities associated with the construction of fissile-material production facilities would have been necessary. In fact, all the associated resource demand components relating to the production of fissile materials would have been satisfied—thirty-three years before the program was begun! Countries with advanced nuclear-related infrastructures should be able to produce nuclear weapons in less than two years (Wohlstetter et al. 1979, 42-46).

From the associated program activities, then, one can work backward to devise a series of indicators that maximizes the probability of being able to carry out particular activities. Placing heavy emphasis on the "Sears, Roebuck" factor and the "Yellow Pages" factor, the resultant indicators for the base case program span a range from indigenous chemical, metallurgical, and electronics production capabilities (including engineering personnel) to construction work force characteristics. (The derivation of the technical indicators is found in Appendix B.) Keep in mind that these indicators do not directly measure a country's capability to produce nuclear weapons. Rather, they gauge the general state of the national resource base in terms of the kinds of technical and industrial activities that would be demanded by a nuclear weapons program. The indicators try to provide, in a systematic fashion, some overall sense of the technical-industrial capacity of the nations. In this regard no single indicator—or set of indicators—is that informative. Here the whole is truly greater than the sum of its parts. A country that successfully scores on *all* the indicators can be judged as having a high probability of being able to undertake each of the activities of the postulated base case program.

ECONOMIC RESOURCE CAPABILITIES AND PROGRAM COSTS

The importance of the economic resource demand components goes beyond the need to consider the monetary costs of a nuclear weapons program. In particular, the total cost of a nuclear weapons program, compared with a more general indicator of a nation's economic capacity, should reflect to a considerable degree the extent to which such a program will be a burden to the national economy, and therefore to the national resource base in general. In other words, while the scientific, technological, and industrial resource indicators described previously allow for inferences regarding national abilities to undertake the *individual activities* contained in a nuclear weapons program, the economic resource indicators discussed here should permit an evaluation of relative national abilities to pursue such activities *as a coherent program*. In this respect a comparison between the total cost of the base case nuclear weapons program and indicators of national economic performance also can be thought of as a measure of administrative burden.

Costs for the base case nuclear weapons program were estimated in an item-by-item analysis (see Appendix C). The costing method reflects changes in information available (e.g., the effect of the Atoms for Peace Program), changing component costs, and so on. The estimated cost for the total base case nuclear weapons program is $61 million in 1960 dollars. It is interesting that in 1960 John McCone, then director of the United States Atomic Energy Commission, observed that "it is possible for a country to develop a plutonium production capability to produce one crude weapon per year with an investment cost of the order of $50 million" (Beaton and Maddox 1962, 22). More contemporary estimates place the six-year cost between $150 million and $200 million in 1980 dollars (Wohlstetter et al. 1979, 45; Office of Technology Assessment 1978)—not far from the corresponding estimate of $172 million derived for this study.

The fraction of national economic resources available for "diversion" to support a nuclear weapons program will depend on the degree of additional burden the nation's decision makers are willing to (have the nation) bear.[15] For example, for military spending in general, data compilations reveal considerable variance in the fraction of national economic product that nations devote to defense. In 1981 about eight countries allocated over 10 percent of their gross national products (GNPs) to defense spending, while twenty-two countries allocated under 3 percent of their respective GNPs (International Institute for Strategic Studies 1981).

What empirical evidence exists actually makes a strong case for the assertion that nations tend not to allocate more than the equivalent of 3 percent of their annual military spending for *initial* nuclear weapons production.[16] If 3 percent seems uncomfortably low, keep in mind that this applies to the *initial phase* of the nuclear weapons program—not the later phases that involve the evolution of a more substantial nuclear force. It excludes all costs related to delivery systems and command and control. The point is that, contrary to popular belief, the "buy in" cost (with respect to established levels of annual defense spending) for an initial nuclear weapons production capability has historically been rather low. To be sure, one can conceive of a number of explanations for these extremely modest initial commitments to nuclear weaponry: the need to maintain existing conventional military strength while national nuclear strength develops, decision makers' natural aversion to placing too many "defense eggs" in a single new "weapons technology basket," the lack of an extensive organizational subgroup with a stake in nuclear weaponry (as opposed to the bureaucratic power of, say, the national tank command, or artillery command), and so forth. Then too, we are talking about 3 percent of what is in most cases a very large number. For instance, if the total cost for the base case nuclear weapons program is about $210 million in 1982, it translates to an annual expenditure of about $35 million a year over the six-year program period. Thus a nation with an average annual military budget of $1.2 billion could "buy" the base case nuclear weapons program without crossing the 3 percent level. Some forty or so nations currently spend $1.2 billion a year or more on defense.

The effects of nuclear infrastructures are particularly noteworthy when we consider nuclear weapons costs. Many students of nuclear proliferation fear that prospective nth countries, which are receiving international loans and credits to build up their nuclear infrastructures, are in fact having their future nuclear weapons programs subsidized. To be sure, there can be no denying that any significant nuclear-related infrastructure does substantially reduce nuclear weapons program costs. In 1982 the base case nuclear weapons program cost would drop from $210 million to $70 million for a country with a moderate nuclear infrastructure, and to about $25 million for one with an advanced nuclear infrastructure.

Obviously, nations like the Federal Republic of Germany and India, with extensive nuclear-related infrastructures—specifically, civil nuclear power plant facilities—would not have to contemplate such extensive resource allocations in "going nuclear," especially with respect to capital resources.

NUCLEAR CAPABLE NATIONS

When the technical and economic indicators for each country are examined together, the result is a list of nations with latent capacities and the corresponding dates when they first acquired that basic capability to build nuclear weapons. Simply stated, a country was recorded as first acquiring a latent capacity during the year in which

TABLE 2

DATES WHEN NATIONS FIRST ACQUIRED LATENT CAPABILITIES TO
MANUFACTURE NUCLEAR WEAPONS (1940-82)

Country	Year
Nazi Germany	(1940)[a]
Imperial Japan	(1941)[a]
United States	1942
United Kingdom	1942/1947[b]
Soviet Union	1942
Canada	1948
France	1948
Federal Republic of Germany	1955
Italy	1956
Czechoslovakia	1957
German Democratic Republic	1957
Sweden	1957
People's Republic of China	(1957) 1961[b]
India	1958
Netherlands	1958
Poland	1958
Japan	1959
Australia	1959
Hungary	1961
Brazil	1963
Belgium	1964
Rumania	1965
Norway	1965
Switzerland	1965
Yugoslavia	1966
Israel	1968
Argentina	1968
Spain	1968
Egypt	1969
Taiwan	1969
Turkey	1971
South Korea	1972
Republic of South Africa	1975
Bulgaria	1975
Greece	1979
Austria	1980

[a]Nonestimated implicit dates, based on scientific feasibility research.
[b]The divided cell reflects two independent British efforts.
[c]First date considers Soviet assistance; the second is based on independent Chinese effort.

it satisfied all the technical and economic indicator thresholds (as detailed in Appendixes B and C). This list is given in table 2.

A "face validity" check against similar tabulations and case studies of other researchers reveals a close correspondence (Meyer 1978d). On average, the differences between estimates are of the order of two years (plus or minus). While the dates for European states show the greatest overall agreement, my estimates for Third World states tend to be earlier than estimates found elsewhere in the literature. This might be because the literature displays a technological bias against non-European states and because much more ambitious nuclear weapons programs are often assumed in the literature. Since the method used to generate table 2 ensures that all countries are judged by the same technological and industrial yardstick, and because none of the resultant dates appear too far from the "conventional wisdom," there is good reason to accept them as at least as accurate as prior estimates, with much greater internal consistency.

Summary

Considerable definitional and analytic work was required to develop the technical model used to determine nation-specific dates of acquiring latent capacities. Between 1940 and 1982 some thirty-six countries were, at one time or another, capable of undertaking the manufacture of nuclear weapons. Figure 4 is a time-series plot of the

Fig. 4. Cumulative number of countries with latent capacities.

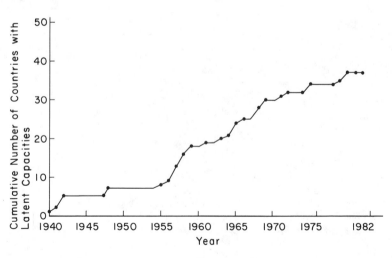

cumulative number of nations with latent capacities (across all nuclear infrastructure categories). As can be seen, during the first fifteen years of the nuclear age the growth in the number of countries with latent capacities averaged about one addition every two years. Between 1955 and 1970 this process accelerated by a factor of three, to about three additions every two years. This acceleration was directly related both to the general emphasis on industrial and economic development that took place after the Second World War and to the more specific international programs aimed at sharing nuclear science and technology. Over the past decade, the rate at which new countries have acquired latent capacities has slowed to about 1.5 every two years. This trailing off is probably because the gap in industrial development between those countries currently with latent capacities and the next group of "candidates" is considerably wider than has been true in the past. In the near term, we may be reaching a saturation point where few additional states can be expected to acquire latent capacities, or at least where the rate of growth of new latent capacities will be much less than it was between 1955 and 1970.

This technical modeling, then, provides the foundation for de-tailed examinations of the technologial imperative hypothesis and the motivational hypothesis. In the former the technological capabil-ity to manufacture nuclear weapons is a sufficient condition for proliferation decisions, while in the latter it plays an important part in the research design. The motivational hypothesis also requires the presence of specific constellations of political military variables. It is the second set of indicators that I shall now develop.

3
The Motivational Basis of Nuclear Proliferation

In composing the array of motivational variables, I have relied on both historical evidence and informed speculation found in the work of other researchers. The historical record provides two complementary streams of information. On the one hand, the "nuclear histories" of the recognized nuclear weapons countries detail the various motive conditions that reportedly led those nations to undertake the manufacture of nuclear weaponry. On the other hand, there are also histories of countries that contemplated but rejected the nuclear option, thus providing information about motive conditions that prompted consideration of the nuclear option but did not result in its selection. Recorded in a similar fashion are the various dissuasive conditions, alleged to have influenced individual nations not to pursue the nuclear option.

The nuclear era, however, is only about forty years old—suggesting that some conditions of potential consequence to the dynamics of nuclear proliferation have yet to evolve. Here the historical record was of little direct aid and it was also necessary to look for guidance to informed opinion and speculation (including pronouncements by foreign policy elites regarding *hypothetical* circumstances that might drive them to produce nuclear weapons.

Again, the motivational data used in this study are not new, but are simply a compilation and extrapolation of the diverse histories found in the literature—albeit taken to a higher level of abstraction.

INDICATOR CONSTRUCTION

As in determining the technical capacity of nations, here too data making offered the opportunity for both subjective assessments and

44

the construction of objective indicators. Particularly important were the complementary strengths and weaknesses of the two approaches. Perhaps even more than for technical assessments, expert evaluations of the politicomilitary perceptions of foreign leaders are clearly susceptible to idiosyncratic biases and prejudices. Consider the recent evaluations of Israeli motives after the raid on Iraq's nuclear center. Some specialists on Middle East politics (with admitted pro-Israeli leanings) have argued that the Israelis did indeed perceive a clear and present danger in Iraq's nuclear development. Other experts (with acknowledged Arabist tendencies) have claimed that Israel's leaders certainly must have been aware of the secret safeguards agreement between Iraq and France and thus could not have perceived any real threat. The rest of the experts (with no clear prejudices) split down the middle.

In contrast, while objective indicators may have biases, they tend to be systematic, correctable, and unrelated to politics. Yet, where objective indicators work well when dealing with well-defined technical variables, the comparatively loose working definitions of political variables make indicator construction that much more difficult and unreliable. To some extent improvements can be made by testing the face validity of the indicators, making explicit comparisons with subjective assessments and historical interpretations found in the literature. This permits the great richness of historical case studies to affect indicator construction directly while the transparency of the objective indicators exposes subjective inconsistency. In this manner the biases and errors of both approaches can be used to reduce the overall error in the final form of the motivational data.

The indicator construction effort leans heavily on two areas of research. As already noted, the selection of incentives and disincentives—and other attendant motive and dissuasive conditions—is based on the theoretical writings and historical research that define the nuclear proliferation literature to date. However, this body of literature could not provide complete guidance in making the indicators operational. In several significant instances I called upon the findings of contemporary world politics research to provide empirical guidance for constructing indicators in particular ways.

To make the most of the many sources of information available, I followed a three-stage strategy for data acquisition. The first stage was to assemble a list of generalized (i.e., abstracted) motive and dissuasive conditions. It seemed that the most efficient approach was to begin with a survey of relevant theoretical literature. (Such works do, after all, lean heavily on history as well as informed speculation.) However, that task turned out to be more complicated than I expected, owing to a terminological problem (noted in chap. 1) involving the haphazard use of the words *motivations, incentives, disincentives, aids,* and *constraints* (*restraints*). To circumvent this problem I had to translate the various discussions into incentive/ disincentive-based hypotheses, thereby providing a common basis for comparison. These, in turn, I then reduced to abstract conditions.

Once I completed that task, it was possible to move to the second stage of developing operational indicators for the individual conditions. The validity of each indicator was then tested against the case study data collected during a survey of historical materials. That validity check was the third stage of the data acquisition strategy. (The validity tests were simple correspondence tests: How closely did the operational results agree with the historical narratives?)

The Motive Conditions for Acquiring Nuclear Weaponry

From the perspective of the motivational hypothesis, decisions to initiate nuclear weapons programs can be understood in the context of three basic categories of incentives: international political power/ prestige incentives, military/security incentives, and domestic political incentives. The first four motive conditions discussed in this chapter are commonly identified with political power/prestige incentives. Under the "influence" of these particular motive conditions, a country may pursue its nuclear option to enhance its status and position in the eyes of other countries. Whether within the context of an alliance or a region, or at the global level, here the incentives to become a nuclear weapons country stem from the belief that such weapons somehow magnify a nation's image. In this respect the extent to which possessing nuclear weaponry actually does enhance a nation's image may be less important than what the prospective nth country believes is true. For global power pretenders eyeing permanent membership on the United Nations Security Council, or a pariah country isolated on the fringe of international activity, the *apparent* utility of nuclear weaponry may be substantial.

The second set of motive conditions is associated with military/ security incentives. Confronted with a military threat from one or more foreign powers, the prospective nth country might turn to the nuclear option in the hope of bolstering its military capabilities. Whether for actual war use (e.g., tactical nuclear weapons) or for deterrence, the acquisition of nuclear weapons may *seem* to represent a viable answer to a variety of military threats (Dunn and Kahn 1976; Gallois 1961; Geneste 1976; Harkabi 1966; Kemp 1974; Kissinger 1957).

The third group of motive conditions is most closely associated with incentives derived from domestic political considerations. That is to say, the decision stimulus originates within the domestic context—with the launching of the nuclear weapons program intended to affect internal, not external conditions. Thus nuclear weapons become a form of domestic political currency. In this respect we must recognize that—to some extent—the effects of *all* the motive conditions will be filtered through the domestic political system *before* a policy decision is made. Therefore all the motive conditions are in some way tied to domestic politics.

The proliferation-related incentives extracted from the theoretical literature are recorded in table 3. The list of incentives was developed by scanning the literature with the following incomplete hypothesis:

> From a decision making perspective, the possession of atomic weapons could be helpful if the government wished to . . .

Note that when the various discussions in the literature are raised one level of generality the result is a finite list of hypothetical incentives, as reflected in table 3. Furthermore, it is important to recognize that the frequency "holes" in the table *cannot* be used to gauge the relative importance of the various incentives, since some authors raised only considerations directly related to the substance of their overall thesis. A further check was also made against the case study data. No "new" incentives were discovered. However, some listed incentives subsume incentives that other researchers discussed individually but that for my purposes represent minor variants.

The next step, then, was to translate the incentives into motive conditions. Since by definition the incentives were condition-derivative, I was able to work backward to determine the underlying motive conditions to which particular incentives were attached. In some instances it was possible to collapse several incentives into a single motive condition. Here I present only abbreviated arguments regarding the theoretical links between motive conditions and decisions to acquire nuclear weapons. Since this study explicitly builds on work by other researchers, the reader is encouraged to consult the sources in table 3.

TABLE 3
A Literature Survey of Proliferation-Related Incentives

Incentive	Beaton and Maddox 1962	Rosecrance 1964	Beaton 1966	Quester 1973	Willrich and Taylor 1974	Dunn and Kahn 1976	Epstein 1977	Greenwood 1977	Office of Technology Assessment 1977	Jensen 1974	Potter 1982
Deter attack from a nuclear armed adversary	X	X	X	X	X	X	X	X	X		X
Redress conventional military Asymmetry	X	X	X	X	X	X	X	X	X		X
Seek military superiority		X				X	X				
Go nuclear before rival				X	X	X	X	X			
Intimidate nonnuclear rivals						X		X			X
Rise to global power status	X	X	X	X	X	X	X	X		X	
Rise to regional power status	X			X		X	X	X			
Enhance general international status					X	X	X	X	X		X
Acquire position in international forums	X		X		X	X	X		X		
Demonstrate national viability (pariah countries)						X	X	X			
Enhance bargaining position within alliance with nuclear powers	X	X	X			X			X		X
Assert political and military independence	X	X	X		X		X		X		X

	1	2	3	4	5	6
Demonstrate modernity (technical prestige)				X	X	X
Economic/industrial spinoffs	X	X		X	X	X
Increase military/scientific morale (most modern weapons)	X	X	X	X	X	X
Increase domestic morale			X			
Divert domestic attention			X		X	
Deter regional intervention by superpower				X		X
Keep up with proliferation trend (regional and/or global)	X	X	X			
Reduce economic defense burden (more bang for buck)	X			X		

[a] 1 = North America, 2 = Latin America, 3 = Europe, 4 = Africa, 5 = Middle East, 6 = Asia.

[b] Beginning date of 1940 corresponds to the first year of this study's temporal domain.

Regional Power Status/Pretensions

A nation's ambition to become or to be acknowledged as a regional power is often mentioned as one of the basic prestige incentives for acquiring nuclear weapons. Yet the argument that links regional power ambitions with the possession of nuclear weaponry may be more subtle than we realize. Even a cursory examination of available sources reveals that nuclear weapons *are not* looked upon as the basis for *achieving* regional power status. The simplistic notion that merely possessing nuclear weapons can transform weak nations into great powers has always been mostly mythology (or wishful thinking). Rather, the argument most commonly encountered is that the possession of nuclear weaponry is *consistent* with regional power status. In other words nuclear weapons are not a substitute for the fundamental bases of regional power but rather a complement to them. "Acquiring nuclear weapons is . . . an act of 'arriving on the nuclear front' as one has arrived, or is arriving, on other fronts of national power and success" (Greenwood et al. 1977, 16). Following this line of reasoning, *regional power status or pretensions* is a motive condition for pursuing the nuclear option.[1]

Regional power status recognizes a particular nation's pivotal role in political, military, and economic relations within the regional setting. Regional power pretensions therefore imply that a nation possesses some "undeveloped" attribute through which it might rise above its regional neighbors. In this respect, a nation's population size compared with its neighbors reflects the relative size of its potential economic market—not to mention its manpower base for military mobilization. The land area of a nation suggests a certain probability of having valuable raw materials, natural resources, and room for expansion (or in a military context, strategic retreat), as well as its geopolitical importance in the region. And, finally, a nation's GNP denotes not only its current economic position but also the size of the economic base for a rapid increase in defense spending.[2]

Taking these three attributes into account, we can devise a simple indicator of regional power status/pretensions. A nation can be considered as having regional power status, or pretensions, if it satisfies two criteria. First, to be considered a regional power a nation must rank among the top three regional countries on any one of the three attributes for at least ten consecutive years during 1940-82.[3] Second, to be considered a regional power in any particular year, it must rank among the top three regional countries on two of three attributes in that year. The first criterion acts more or less as a noise filter, removing those cases where, for anomalous reasons, a nation appears among the top-ranked nations in its region for a short time. The second criterion, then, is the more significant. The resulting list is displayed in table 4.

TABLE 4
REGIONAL POWERS AND REGIONAL POWER PRETENDERS: 1940-82

Country (Region)[a]	Years (inclusive)[b]
United States (1)	1940-
Canada (1)	1940-
Mexico (2)	1940-
Argentina (2)	1940-
Brazil (2)	1940-
Britain (3)	1940-
France (3)	1945-
Germany (3)	1940-45
Federal Republic of Germany (3)	1955-
Italy (3)	1940-43
Soviet Union (3)	1940-
Nigeria (4)	1960-
Ethiopia (4)	1946-
Republic of South Africa (4)	1946-
Iran (5)	1980-
Iraq (5)	1977-
Egypt (5)	1951-73
Saudi Arabia (5)	1974-
China (6)	1946-49
People's Republic of China (6)	1950-
Japan (6)	1940-45
Japan (6)	1952-
India (6)	1947-
Australia (6)	1940-

[a]1 = North America, 2 = Latin America, 3 = Europe, 4 = Africa 5 = Middle East, 6 = Asia.
[b]Beginning date of 1940 corresponds to the first year of this study's temporal domain.

Global Power Status/Pretensions

In an age of superpowers, the concept of global powers (or major powers) may seem outdated. For many the Suez crisis of 1956 represented a watershed—the confluence of superpower interests dictating the flow of international events. Nevertheless, there is a class of nations whose ability to systematically influence events outside their own regions implies a rank greater than "regional" status—though certainly not on a par with those of the superpowers.

51

These nations do seem to have pivotal roles in international relations.

Perhaps the distinction between nations that are global powers and those that are merely "influential" is to be found in the variety of mediums through which the former exert their influence. Global powers may choose to operate in the political sphere, the military sphere, and/or the economic sphere. Other countries, to the extent that they can affect events outside their own regions, are usually limited to a single medium. Thus France and Britain can be considered global powers, given the political (diplomatic), military, and economic tools at their disposal and their ability to employ those tools around the world. But Saudi Arabia and Iran (before the revolution), whose abilities are confined to the economic sphere of influence, cannot be properly viewed as global powers. In particular, their military and political abilities are limited in scope.

As is true for regional power status, nuclear weapons have historically been perceived as consonant with global power status— indeed, even more so.[4] The nuclear status of the permanent members of the United Nations Security Council is perhaps the obvious illustration. For countries with global power pretensions the illustration is well taken, since it does indeed appear that possessing nuclear weaponry is a necessary—if not sufficient—condition for permanent membership on the council. Thus, *global power status/ pretensions could easily provide the motivation for a nation to pursue its nuclear option.*

Toward the objective of identifying the global powers, procedures for developing objective indicators of "global power" status have been argued and compared in a number of scholarly works.[5] Conveniently, most of the resultant lists tend to converge at the top; though their internal rankings may differ, their "top ten" compositions follow historical consensus. This convergence in the composition of the top ten probably occurs because there is a general agreement on the underlying bases (indicators) of global power status. The disagreement about relative rankings can probably be attributed to different methods of combining those indicators into a single index. The latter consideration, however, is not relevant to this study. Accordingly, table 5 presents the list of global powers used here.[6]

Somewhat more problematic was how to go about specifying the crop of global power "contenders." Not only is a list necessary, but it is equally important to have some idea of the point when these countries seriously began to see themselves as global power contenders (see, e.g., Brown 1974; Spiegel 1972). The approach em-

TABLE 5
GLOBAL POWERS: 1940-82

Country	Years (inclusive)[a]
United States	1940-
Soviet Union	1940-
Britain	1940-
France	1945-
Germany	1940-45
Japan	1940-45
Italy	1940-43
People's Republic of China	1950-

[a]Beginning dates of 1940 correspond to the first year of this study's temporal domain. For most of these nations their major power status begins much earlier. The dates used here are found in Singer and Small (1972).

ployed is admittedly less operational than I would prefer. It does, however, have considerable intuitive appeal. The nations selected were those most often mentioned in the literature on world politics as near-future prospects for global power status. Not surprisingly, all the candidates were drawn from the regional powers list. Corresponding inclusion dates were based on particularly noteworthy events that would have focused domestic and government attention on the nations' global power potential.

Owing to their industrial/economic strength, both Japan and the Federal Republic of Germany have been mentioned frequently as potential global powers. Certainly their levels of technology and industrial production are equivalent to, if not greater than, those of the recognized global powers. Taking this as a cue, I postulated that once these nations passed the two most comparable global powers— Britain and France—in their GNPs, the issue of their global power status would be sure to arise in international discussions. But we are really concerned with the specific nation's self-image. Accordingly, for the Federal Republic of Germany this corresponds to 1958, while for Japan the date is about 1965.

Also of interest were the two most often noted "nontraditional" contenders, India and Brazil. Their claims to global power status rest primarily on their vast untapped resources—notably population and territory—as well as their "strategic" locations. India has certainly not tried to disguise its global power pretensions. Since the early 1960s a number of government officials, scholars, and intellectuals have referred to India's great power destiny (Gupta 1965, 59-70; Subrahmanyam 1974, 122, 127, 135). But such an eventuality had always been discussed in terms of an undefined time horizon. However, political and military events between late 1971 and late 1972 seem to have brought India's great power ambitions one step closer to reality (Bhargava 1976). Among the most important events was the elimination of Pakistan as a serious regional rival following the Bangladesh War of 1971. Pakistan's seemingly permanent dismemberment by early 1972 marked India's emergence as a dominant power in South Asia (Bhargava 1976, 3). Thus 1972 was recorded as the beginning date of serious Indian global power pretensions.

In terms of global power aspirations, Brazil and India are very similar. With the establishment of a military government in Brazil in 1964, the possibility of mobilizing Brazil's vast resources toward achieving global power status received increasing attention (Gall 1976, 177-78). The subsequent and much heralded expansion of the Brazilian economy throughout the late 1960s and early 1970s focused international attention on the Latin American giant. No doubt it also led Brazilians to assess their own potential. It was also during this period that Brazil began more active diplomatic involvement, especially in the context of the NPT debates (Epstein 1976; Quester 1973; Stockholm International Peace Research Institute 1972). Other indications of the evolution of a truly independent Brazilian foreign policy included Brazil's United Nations vote to condemn Israel (1975) and its early recognition of the Marxist government of Angola.[7] But if one were to search for a high-water mark, it would be the period surrounding the successful conclusion of the Brazilian-German deal for nuclear technology transfers (February 1975). Statements by Brazilian officials reveal a strong linkage between the nuclear energy independence signified by the agreement and a Brazilian self-image of global power status (Gall 1976; Office of Technology Assessment 1977, vol. 2, app. 1). Apparently, from the Brazilian perspective the acquisition of a complete nuclear fuel cycle came to be equated with national technological and industrial advancement. Of special significance was the visit by United States Secretary of State Kissinger to Brazil in February 1976. Observed Kissinger: "Brazil's diplomats speak for a nation of greatness—a people taking their place in the front rank of nations, a country of continental proportions . . . a nation now playing a role in the world commensurate with its great history and its even greater promise" (*New York Times* 1976a, 3). If the Brazilians had any lingering doubts about their potential to be a global power, the Kissinger "baptism" most certainly eliminated them. Furthermore, the United States and Brazil went on to sign an agreement pledging regular consultations between the two nations on issues of global import.[8]

It seems clear then that a series of important changes in Brazil's self-image took place between 1974 and 1976. Following an averaging strategy, I coded 1975 as the beginning date for serious Brazilian pretensions.

Pariah Status

There is a small group of nations that can be characterized as international pariahs, or outcasts. These are countries that for one reason or another have been shunned by their regional neighbors, if not by the international community in general (Chan 1980; Harkavy 1977; Betts 1977). The logic of this situation suggests a need to demonstrate national viability. In this context some scholars have suggested that countries with pariah status might pursue the nuclear option to force the international community to sit up and take note. The pariah's acquisition of atomic weapons would make it impossible for the countries of the world, in particular the regional countries, to continue to ignore it.

As a first cut, a country was coded as having pariah status if it did not have diplomatic relations with at least 10 percent of its regional neighbors. Diplomatic relations were indicated by the presence of foreign consulates or embassies. Unfortunately, the validity check revealed that under that rule more than half the countries of Africa appeared as pariah nations. (The relatively high cost of maintaining foreign embassies apparently prevented many new nations from establishing formal ambassadorial posts.) Therefore two further criteria were appended. A country was coded as having pariah status if by five years after achieving independence it did not have diplomatic relations with at least 10 percent of its regional neighbors and belong to at least one regional alliance/organization.

When the validity check was performed the second time, the results confirmed a high degree of (judgmental) historical correspondence. The resulting list of pariah countries is presented in table 6.

A Military Alliance with a Nuclear Weapons Power

Here two proliferation-related incentives are superimposed: a nation's desire to enhance its bargaining position within an alliance with a nuclear power, and its desire to assert politico-military independence. In both instances the underlying conditions are the same. A country perceives itself to be in an inferior position within an alliance structure. In particular, the dominant power is a nuclear power. All else equal, the acquisition of atomic weapons would

TABLE 6
INTERNATIONAL PARIAHS

Country	Years (inclusive)
Cuba	1970-
Israel	1955-
Republic of South Africa	1965-
South Korea	1955-59[a]
Taiwan	1975-

[a]In 1960 the number of Asian nations having diplomatic relations with South Korea climbed above 10 percent. Therefore South Korea "lost" its pariah status.

theoretically increase the military significance of the weaker partner—thereby enhancing its status within the alliance.

This motive condition also has been considered by some to act dissuasively, providing disincentives against proliferation. This second aspect is discussed below.

The operationalizing of this motive condition is self-evident: countries with a formal defense pact with a nuclear weapons power were recorded as having a "nuclear ally."[9]

A Security Threat from a Nuclear-Armed Adversary

The first incentive noted in table 3, deterring an attack from a nuclear-armed adversary, refers to both a conventional attack and a nuclear attack. In essence the potential proliferant has sufficient reason to suspect it *could* become involved in a security dispute with an existing nuclear weapons country. India (1964-74), vis-à-vis the People's Republic of China, Pakistan (post-1974) vis-à-vis India, and the Soviet Union (1945-49) vis-à-vis the United States are all appropriate illustrations. Obviously the specification of this motive condition is based on two independent criteria: (1) the potential nth country perceives some likelihood of future security disputes with a specific country; and (2) that potential adversary is known to possess atomic weapons. Thus two distinct steps were required to make this variable operational. First I determined potential adversary dyads, then I made a straightforward determination whether either (or both) of the dyad members was a recognized nuclear weapons country.

From the perspective of the realpolitik literature, interstate adversary relationships may be either explicit or implicit (Kissinger 1964; Morgenthau 1973; Organski 1968; Spykman 1942; Wolfers 1962). An explicit adversary relationship exists if two nations have had a recent security dispute—the Indian/Pakistani adversary dyad is a good illustration. Expectations of a future dispute are founded on precedent. Israel and the Arab states is another obvious example. An implicit adversary relationship, however, lacks a behavioral referent. Rather, expectations of a future dispute are premised on potentially contentious interactions resulting from specifically ascribed roles. In other words, though two countries may not have a recent history of mutually disruptive behavior, their ascribed roles imply the prospect of conflict. Brazil and Argentina, for instance, have not had a security dispute in the post-World War II period. Yet most scholars recognize their implicit adversary relationship, the product of their regional power roles. Again, explicit adversary dyads are behavior dependent; implicit adversary dyads are role dependent.

Following this definition, explicit adversary dyads were determined by prior security dispute interactions. The dispute data were coded to record the disputants, the beginning year (and, if there was one, the terminating year) of the dispute, and the aggregate number of casualties.[10] (Where multiparty disputes were encountered, I used historical descriptions to disaggregate the data into primary adversary dyads.)

I performed two analogous data transformations. Working under the assumption that conflictual behavior (as denoted by the coded security disputes) does not materialize spontaneously, I coded the first year of the adversary dyad as the year before the beginning of the observed security dispute. Here I presumed that the nations in an adversary dyad are aware of their rivalry before it surfaces in behavior. Similarly, to allow for the persisting "aftereffects" of a terminated security dispute I used two transformations: (1) if the dispute did not involve a war, its aftereffects were presumed to dissipate after five years; (2) if the dispute did involve a war, its aftereffects were presumed to persist for ten years after a peace treaty was signed. If no treaty was signed, the adversary relationship was presumed to continue.[11, 12] Thus the United States/Soviet Cuban missile crisis (1962) reflected a United States/Soviet adversary dyad from 1961 to 1967. The 1967 Arab/Israeli War, representing several dispute dyads, continues today.

In specifying the implicit adversary dyads I leaned heavily on the world politics literature for theoretical guidance. First, for each year I partitioned the membership of the international system into five distinct groups: global powers, regional powers, alliance partners, declared neutrals, and others.[13] The *global power group* consisted of those countries recognized as having a key voice in international political and military affairs. They are explicitly concerned with exercising their influence in several corners of the globe, and they expect to be consulted about every issue of consequence. The *regional power group* is conceptually very similar to the global power group, except that their fundamental interests (and their acknowledged status) are confined to a semidefined geopolitical area. From the perspective of hegemony, regional powers see themselves as deserving a special voice in regional issues. The *alliance partner* designation denotes a nation whose participation within a defense coalition implies an adversary relationship with the target(s) of that alliance. For example, Iran's participation in CENTO was assumed to reflect an implicit adversary relationship with the Soviet Union.[14] In contrast to the alliance partner group, the *declared neutrals* are those (few) countries that have stated their intention to remain, and are generally recognized as existing, outside *any* alliance or alignment arrangement.[15] Quite simply, the declared neutrals are Sweden, Switzerland, and Austria. Finally, the "others" group represents a residual category that corresponds to those nations whose primary concerns center on local issues (Spiegel 1972). Next, I had to devise an algorithm for determining adversary dyads as a function of ascribed role. If the world politics literature converges on any single point, it is that global power relations can be characterized as competitive and contentious. In a sense global powers—by virtue of their status—are natural security threats to each other. Therefore I assumed that all (nonallied) global powers would tend to view each other as adversaries. Similarly, the dynamics of regional power "mentalities" and behaviors were assumed to mimic those of global powers, but only within the geopolitical setting of the region. As already noted, where alliances were known to have a specific target, I coded implicit adversary dyads accordingly.

For declared neutrals I followed a rather simplistic rule. Because by definition declared neutrals cannot enter into alliances, an implicit adversary relationship involving the neutral country was assumed to exist with all regional powers and all regionally allied global powers.[16] This rule tends to distort the neutrals' security setting, yet it does capture the "go it alone" aspect of neutral status.

Finally, as I have noted, the "other" category connotes countries whose primary concerns center on local issues. These nations have no generic "role" from which I could make inferences regarding implicit adversary relationships. Indeed, any implicit adversary relationships are likely to be highly specific (idiosyncratic). Therefore I assumed that these countries' overall adversary situations would be reflected by explicit adversary relationships alone.

Almost by definition, then, we arrive at the following algorithms:

Global powers are coded as facing a security threat from a nuclear-armed adversary if any global power (excluding defense pact allies), any "local" regional powers (excluding defense pact allies), or any other country with which it has had a conflict-prone relationship possesses nuclear weapons.

Regional powers are coded as facing a security threat from a nuclear-armed adversary if any regionally allied global power (excluding defense pact allies), a "local" regional power (excluding defense pact allies), or any other country with which it has had a conflict-prone relationship possesses atomic weapons.

Alliance partners are coded as facing a security threat from a nuclear-armed adversary if the explicit target(s) of the alliance (if there was one), or any other country with which it has had a conflict-prone relationship possesses atomic weapons.

Declared neutrals are coded as facing a security threat from a nuclear-armed adversary if any regional power, any regionally allied global power, or any other country with which it has had a conflict-prone relationship possesses atomic weapons.

Those countries not included in any of the groups above are coded as facing a security threat from a nuclear-armed adversary if any country with which they have had a conflict-prone relationship possesses atomic weapons.

An Adversary with a Latent Capacity

The avoidance of risk (hedging against uncertainty) in national security planning has been institutionalized through worst-case planning (Garthoff 1978). A country that believes an adversary nation is capable of manufacturing nuclear weapons may assume the worst: that its rival is indeed about to produce, or is already secretly producing, nuclear weapons. Israel's suspicion of Iraqi intentions comes readily to mind. One obvious response is to initiate one's own nuclear weapons program. The incentive here would be to beat one's rival to the punch—to get a head start on the production of nuclear weapons. (Some have suggested that this variable may also exert a dissuasive influence; I will discuss this aspect later.)

The algorithm for making this condition operational followed the format outlined above for nuclear adversaries, except that the adversary nation did not have to possess nuclear weapons, but merely required a latent capacity to produce them. Finally, to provide for "anticipatory" fears, I used a three-year lead factor, reflecting one-half the time needed to develop an operational capability when starting with a latent capacity (but without any nuclear infrastructure). Here the lead factor is used in recognition that a nation may first have to make a conscious effort to acquire a latent capacity. Pakistan and Iraq are good examples. Other countries, in turn, may notice this effort and respond with their own nuclear weapons programs.

An Overwhelming Conventional Military Threat

It is often suggested that countries consider the nuclear option a way to redress an existing conventional military asymmetry in which an adversary country has overwhelming superiority in conventional forces. Viewed in this context, nuclear weapons—with their incredibe destructive power—can be looked upon as "equalizers." To be sure, the overwhelming conventional threat has been given as the military rationale for the deployment of United States tactical nuclear weapons in Europe and East Asia.

Here again the specification of the adversary (or target) country follows the scheme outlined in the section dealing with nuclear adversaries. The operationalizing of an *overwhelming conventional threat*, however, requires some elaboration.

In general, conventional military superiority can arise from a superiority in manpower, a superiority in conventional military equipment (i.e., technological intensity), or both. Manpower comparisons are simple: armed force size versus armed force size. Evaluating the technological intensity of military forces is more problematic. Ideally, we could utilize vast time-series data arrays that reflect not only the quantities of major weapons systems but their relative lethality, vulnerabilities, and so forth. A simple dynamic analysis, comparing aggregate firepower scores, for example, might yield interesting results. While in the long run such a data base might be most desirable, in the near term assembling this data is simply not feasible. For this study, an alternative approach proved sufficient. Assuming that the average defense dollar spent per soldier offers a rough approximation to the general technological intensity of a country's armed forces[17] (but not in a strict metric sense), one way to depict the conventional military threat posed by a target country (T), toward a potential proliferant (P), is:

$$R_{T,P} = \sqrt{\frac{(A)^2 + (M)^2}{(B)^2 + (N)^2}},$$

where

$R_{T,P}$ = an indicator of the conventional military threat that the country poses to the potential proliferant

$$A = \begin{cases} AF_T/AF_P & \text{if } AF_T > AF_P \\ 1 & \text{if } AF_T \leq AF_P \end{cases}$$

$$B = \begin{cases} 1 & \text{if } AF_T \geq AF_P \\ AF_P/AF_T & \text{if } AF_T < AF_P \end{cases}$$

$$M = \begin{cases} MILEXP_T/MILEXP_P & \text{if } MILEXP_T > MILEXP_P \\ 1 & \text{if } MILEXP_T \leq MILEXP_P \end{cases}$$

$$N = \begin{cases} 1 & \text{if } MILEXP_T \geq MILEXP_P \\ MILEXP_P/MILEXP_T & \text{if } MILEXP_T < MILEXP_P \end{cases}$$

and

AF_T = armed force size of target country
AF_P = armed force size of potential proliferant
$MILEXP_T$ = military expenditures of target country
$MILEXP_P$ = military expenditures of potential proliferant.

This simple model treats military expenditures and armed force sizes as independent components that together produce "conventional military power." Thus the resulting force (i.e., conventional military power) can be computed in accordance with ordinary vector calculus (hence the mathematical form of the indicator). In an analytic sense there is some correspondence between this (form of) indicator and the method of comparing conventional military capabilities via static indicators. The primary difference is that the former permits the computation of a single summary index of conventional military power.

Moreover, this particular formulation possesses several interesting properties. First, absolute differences in manpower levels translate directly into proportional absolute differences in indicated conventional military power. Thus, if two countries spent equal defense dollars per soldier but one country had twice as many soldiers, the second would have twice the conventional military power.

Technological intensity, however, does not behave this way. Indeed, incremental increases in weapon effectiveness are often achieved through disproportionate increases in cost. For this reason absolute differences in technological intensity, as sensed by this indicator, are not translated into equivalent military power gains. That is, doubling the technological intensity of one's military forces (doubling military expenditure per man) would not double one's *indicated* level of conventional military power. Rather, the indicated increment in conventional military power will always be less than the corresponding increment in technological intensity. Finally, the indicator discounts small differences in technological intensity, while large differences have a greater effect. Consequently many probable differences between different nations' military expenditures per man, resulting from pay scale differentials, varying factor costs, currency conversion rates, and so forth, should not produce any large systematic bias in the indicator.

What ratio of conventional military power reflects an overwhelming conventional military threat? The conventional wisdom is that *defending* nations (what country would ever admit to being anything other than a defender?) can usually hold their own up to a three-to-one ratio of military force disadvantage (Liddell Hart 1954; Planning Research Corporation 1968). Thus I postulated that an overwhelming conventional military threat would be one in which $_{RT,P} \geq 3.0$ (i.e., three-to-one or greater in favor of the target country.)[18]

The validity check against historical data did not, however, confirm this. Several important cases (e.g., the United Kingdom vis-à-vis the Soviet Union from 1946 to 1951) were not detected by the operational indicator. So the indicator was tested for $_{RT,P} \geq 2.5$. Again, several important cases did not materialize. With $_{RT,P} \geq 2.0$, all the historical referents were properly indicated, but a number of false cases were also generated. For example, Egypt, Syria, Jordan, Taiwan, South Korea, Australia, and most of the membership of NATO and the Warsaw Pact were tagged as facing overwhelming conventional military threats during the late 1950s and through the 1960s.

Examining the false cases revealed an interesting similarity. The reason they did not appear in the historical data was that each possessed an alliance commitment. When treated as isolated countries, each seemed to face an overwhelming conventional military threat. But when viewed as operating within a defense alliance, the aggregate conventional military capabilities of the alliances negated the overwhelming conventional threat. To be sure, alliances are very effective counters to conventional military threats—a point so obvious that I am somewhat embarrassed to admit overlooking it. In correcting for alliance participation, I found that the following algorithm produced results highly convergent with the historical data. First, I recorded all cases in which $_{RT,P} \geq 2.0$. Then I deleted those cases involving nonglobal powers if the nation under consideration had at least one global power alliance partner or three nonglobal power alliance partners. If the nation was itself a global power, then I deleted the case if it had at least three global power alliance commitments (i.e., an alliance with three global powers, three different alliance commitments from the same global power, or any combination totaling three global power alliance commitments).[19]

As it turns out, the alliance formulation used has some intuitive merit. For nonglobal powers, global power alliances offer greater security (credibility) than alliances with other nations. But for global powers the credibility of an alliance with other global powers is always a shaky proposition, given the possibilities for future conflicts of interest (e.g., the "defection" of the United States during the Suez crisis of 1956). Thus, several global power commitments offer greater assurance to the global power itself (Sabrosky 1975).

Keep in mind that this indicator does not attempt to measure the actual level of the conventional military balance between adversaries. Instead, it tries to reflect how that balance is likely to be *perceived* by political decision makers. As the recent defense debate in the United States shows, comparisons of military manpower, military spending, and alliance relations are often the indicators political leaders look toward.

In sum, countries were coded as facing an overwhelming conventional threat if $_{RT,P} \geq 2.0$ and their alliance status did not satisfy the criteria outlined above.

Domestic Turmoil

It has been suggested that national decision makers consider the nuclear option a way to direct domestic energies away from domestic problems. In essence this represents a technological approach to rallying round the flag. Enhancing domestic morale in the face of civil strife, ethnic hostility, and labor unrest by acquiring nuclear weapons is clearly a questionable proposition—though by no means inconceivable.

Domestic unrest is to some extent inherent in most national experience. Consequently one might postulate that a recognizable— and by implication frightening—increase in domestic unrest would be necessary to move national leaders to consider the nuclear option. To make this motive condition operational, therefore, I used the following indicators: (a) general strikes, (b) riots, and (c) antigovernment demonstrations.[20] The annual magnitudes of each of the three indicators were summed, in the sense that they represent discrete events. If in any year the total equaled or exceeded twelve events (signifying an average of at least one major domestic disturbance a month), and if the total was at least double that of the previous year, then the situation was coded as an alarming degree of domestic unrest. Taking into consideration lagged effects, the subsequent year was also coded as exhibiting domestic turmoil.

The validity check proved sufficiently reassuring—though historical references to alarming levels of domestic unrest were relatively few. Still, the resulting index did capture those cases found in the historical survey, while the number of nonreferenced cases was comparatively small (and conceivably correct).

Major Military Defeat and National Self-Image

Though the literature is somewhat ambiguous here, there appears to be a general belief that governments might pursue the manufacture of nuclear weapons as a way to raise the morale of their defense establishments. The only such historical reference to a recognized nuclear weapons country I could find involved the French military in the late 1950s,—specifically, the demoralizing effects of the successive defeats in Indochina and Suez (Kelly 1960, 293; Kohl 1971, 357). Pakistan's response to its dismemberment in 1971 is an example where speculation suggests a relation. Thus it seems plausible that nuclear weaponry, as a symbol of military prowess and advance capabilities, could be viewed as a "pain reliever" for a demoralized defense establishment (Joshua and Hahn 1973, 16), a condition arising from a recent military defeat. The military defeat, then, becomes the actual motive condition. Accordingly, the unsuccessful war experiences of the international membership from 1940 to 1980 were used as the indicator.[21]

Regional Nuclear Proliferation

Even if a new nuclear weapons country is not a rival of any of its neighbors, its action may nonetheless stimulate further regional proliferation. Basically this argument is grounded in three premises: (1) though not a rival today, it could be one in the future (e.g., United States perceptions of the Soviet Union during World War II); (2) an impulse to "keep up with one's neighbors" is created by each new nuclear weapons country; and (3) the emergence of a new nuclear weapons power will compel other nations to reconsider their own situations in light of one more nation's decision to go nuclear (i.e., it raises doubts about prior decisions to forgo the nuclear option).

Operationally, if any nation in a given region provided unambiguous evidence that it possessed atomic weapons in a specific year, then regional nuclear proliferation was coded as having occurred.

Moreover, the "aftereffects" of this event were considered to persist for two years beyond the actual proliferation event. Thus I chose a three-year window for decision because it represents one-half the period needed to implement a nuclear weapons program and because it seemed unlikely that bureaucratic delays in reacting to such an event would extend much beyond three years.

Intolerable Economic Defense Burden

In the context of the economics of modern military forces, one can argue that a nation might "go nuclear" in the hope of getting "more bang for the buck." The proposition is, of course, that once the cost of conventional forces becomes intolerable, acquiring nuclear weapons would permit the nation to maintain (or increase) its military capability with less of an economic burden. Of course such a high defense burden is also an indicator of either perceived security threats or militaristic tendencies. In either case nuclear weapons may be viewed as desirable.

Exactly what constitutes an intolerable defense burden is an empirical question that is difficult to isolate in time or space. The disarmament literature tends to use 10 percent of the GNP as a benchmark (Hitch and McKean 1967; Russett 1970; Weidenbaum 1967). Certainly 10 percent seems to represent a psychological threshold—if not a true burden level. Consequently, I used 10 percent of the GNP as the threshold point. A nation whose defense spending equaled or exceeded 10 percent of its GNP for at least three consecutive years (and displayed a zero or positive trend) was coded as facing an intolerable defense burden for the year beginning the three-year series. The criterion of three consecutive years was adopted to filter out anomalous jumps of the indicator across the threshold point.

Incentives and Motives Indirectly Included or Omitted

Several of the incentives noted in table 3 cannot be tied to any single motive condition—they cut across a number of putative causes of nuclear proliferation. Thus they are treated implicitly rather than explicitly in development of the motivational model. For example, nations seeking military superiority are subsumed within those motive conditions concerned with adversary military capabilities: for example, an adversary with nuclear weaponry, an adversary with a latent capacity to produce nuclear weapons, or an adversary with overwhelming conventional military capabilities. Here too the quest for military superiority may be *simultaneously* reflected by the defense expenditure burden indicator. Beyond this it is difficult to assess, objectively or subjectively, when a nation is seeking military superiority over an adversary for the sake of superiority itself. For nations pursuing military superiority in a more general regional context, regional power pretensions coupled with an indicated defense expenditure burden can be expected to reflect this incentive.

The incentive of deterring regional intervention by a superpower was dealt with in a similar fashion. I assumed it was related to the regional or global power role and therefore implicit in the indicators for region and global power pretensions. In like manner the incentives related to enhancing a nation's general international status, demonstrating modernity, and benefiting from economic/industrial spinoffs all represent universal international desires. All countries should be expected to operate under their influence all the time. So they are particularly difficult to isolate. In one sense, though, one can argue that they are most strongly associated with regional and global power considerations and, as such, are captured by those indicators. The same was assumed to be true of the incentive "acquiring an enhanced position in international forums."

Alliance relations play an important role in several of the indicators, particularly with respect to "overwhelming conventional threat" as a motive condition and "nuclear ally" as a dissuasive condition. Though it is easy to take for granted the "credibility" of alliance relations, there are several well-known instances where the continued existence of formal alliance relations merely papered over the fact that the substance of the alliance relationship had evaporated. The Sino-Soviet "friendship" treaty after 1958 is one example. The United States-Korean alliance during the mid-1970s—undermined by United States declarations of intention to pull out of Korea—was viewed by the Koreans as of questionable credibility. The United States refusal to back France and Britain during the 1956 Suez crisis, and United States reluctance to support Greece in the 1974 Cyprus crisis are also relevant. Similarly, the United States-Taiwan alliance was explicitly called into question in the wake of the Shanghai agreements. We must also include United States-Pakistani alliance relations from 1972 to the present as an example of a breakdown in credibility—owing to the unwillingness or inability of the United States to prevent the dismemberment of Pakistan in 1972. In each case the image and the reality of the alliances were at variance. Thus a potentially significant incentive for acquiring nuclear weapons becomes the desire to redress a perceived change in the politico-military milieu resulting from the loss in credibility of an alliance. (Here the emphasis is on alliances with major powers.) This incentive was incorporated into the data base through its effects on the motive conditions nuclear threat and overwhelming conventional threat and on the dissuasive factor nuclear ally where appropriate. When an alliance existed but its credibility was in question, its formal existence was ignored in specifying nuclear threat, overwhelming conventional threat, and nuclear ally.

Two additional incentives were not included in the study: a general trend toward nuclear proliferation and continued vertical proliferation. In the case of the former, many countries have gone on record as declaring that they will remain nonnuclear weapons countries only as long as the global nonproliferation regime holds up. In other words, if the acquisition of atomic weapons seemed to be becoming the rule rather than the exception, they would feel compelled to follow suit.[22] The problem that arises is that government spokesmen have been particularly careful not to disclose their criteria for determining when nuclear proliferation has crossed the line from exception to rule. Moreover, it is pretty much accepted that to date—regardless of how one defines a general trend toward nuclear proliferation—it has not yet manifested itself. For the purposes of the analysis, then, there is no need to be concerned about creating objective indicators. This incentive does, however, remain a serious issue for forecasting future developments in nuclear proliferation.

The second excluded incentive, also related to a general trend toward nuclear proliferation, emphasizes the vertical aspect—notably, the United States-Soviet nuclear arms buildup. The problem with this incentive is that it is in essence a global constant; since the 1950s vertical proliferation has been fairly steady and ever present. Even during the détente period with its SALT negotiations, both sides continued to add missiles, warheads, and other nuclear weapons to their stockpiles. Because of the constancy of this indicator, its inclusion or exclusion in the analysis will have no measurable effect on the results. Consequently it was not included in the list of motive conditions.

DISSUASIVE FACTORS THAT AFFECT CONSIDERATION OF THE NUCLEAR OPTION

As important as the motive conditions are to the progress of nuclear proliferation, there is a collection of dissuasive conditions that tend to work against the nuclear option. Table 7 lists the dominant dissuasive conditions found in the literature.

TABLE 7
A LITERATURE SURVEY OF PROLIFERATION-RELATED DISSUASIVE CONDITIONS

Incentive	Beaton and Maddox 1962	Rose-crance 1964[a]	Beaton 1966[b]	Wilrich and Taylor 1974	Office of Technology Assessment 1977	Greenwood 1977	Epstein 1977	Dunn and Kahn 1976	Potter 1982
Alliance with nuclear power	X		X			X	X		X
"Peaceful reputation"	X	X		X	X			X	
Becoming nuclear target		X				X	X		X
Diluting alliance bond				X				X	X
International legal commitments		X	X			X	X		X
Domestic politics		X			X				X
Rival with latent capacity				X	X	X		X	
Risk of unauthorized seizure						X		X	
Intervention by nuclear power						X	X		
Technical/economic incapacity	X	X	X	X	X	X	X	X	

An Alliance with a Nuclear Power

One of the more obvious conditions that might negate incentives for nuclear proliferation is an alliance with a nuclear power. In this situation the "client" nation seeks security beneath a nuclear umbrella provided by the nuclear alliance partner. The NATO countries, Japan, and South Korea, for example, expect United States nuclear support. The Eastern European nations expect similar support from the Soviet Union. In a sense the alliance acts as a surrogate for a national nuclear capability.

The notion of diluting an alliance bond can also be considered a disincentive. The basic argument is that an alliance client that "goes nuclear" risks the dissolution of the alliance bonds. The United States, for example, might well abandon its alliance with South Korea in the face of South Korean nuclear proliferation. Thus an alliance with a nuclear power plays a double role in dissuading countries from going nuclear. First, it acts in opposition to several of the motivational conditions that have been noted (e.g., an overwhelming conventional threat). Second, the possibility that the alliance's major partner might be compelled to dilute or dissolve the alliance bond acts to restrain the potential proliferant.[23]

This indicator is, of course, identical to that discussed in the section on motive conditions. Thus it becomes an empirical problem to determine whether one, or both, of these hypothesized traits exists.

International Legal Commitments

Simply stated, a country that has assumed an international legal obligation not to manufacture nuclear weapons faces potentially severe repercussions should it decide to abrogate the agreement and begin producing atomic weapons. On the one hand, the offending nation may have to tolerate the resultant political fallout—diplomatic protests, insinuations regarding its intentions, strains in diplomatic relations, and so forth. Depending upon the particular situation, the country might have to contend with imposed economic, technical, military, and/or trade sanctions. On the other hand, such an action might also stimulate a chain reaction in which regional neighbors, especially rival countries, follow suit.

International legal commitments also may influence bureaucratic bargaining and the balance of pressures within the domestic setting. That is to say, both the composition of the relevant group of decision making actors and their relative influence in the decision process may be affected by nonproliferation treaty commitments.

Of course, a country's assuming such legal commitments may connote the existence of other dissuasive conditions. A legal treaty may serve as a surrogate indicator for other dissuasive conditions that are not directly observable. For instance, a domestic political consensus against becoming a nuclear weapons country, or at least domestic ambivalence, would be a necessary prerequisite for this type of international commitment. The inaccessibility of nuclear technology assistance might also manifest itself in accession to a legal treaty. Libya, for example, is believed to have ratified the NPT at the insistence of the Soviet Union, which otherwise refused to provide nuclear assistance. Consequently, the dissuasive potency of international legal commitments flows from two sources—one, the underlying bases for having assumed the commitment(s), and two, the ex post facto potential repercussions that would follow from abrogating the agreement(s). On the one hand, the latter effects are independent of the former, so that even if a swing in domestic opinion occurs the dissuasive power of the international legal commitment(s) may still be effective. On the other hand, should legal treaty be just a surrogate for one or more dissuasive factors, and should those factors disappear, the dissuasive effects of legal treaty could be small or nil.

An obvious indicator for legal commitments against nuclear proliferation is the year in which a nation assumes an international legal obligation not to manufacture nuclear weapons. The archetypal agreements include the NPT, the Treaty of Tlateloco, and the agreements related to the Western European Union. If a nation acceded to more than one independent agreement, I noted these additional obligations.

Rival with a Latent Capacity

A nation whose rival possesses a latent capacity to manufacture atomic weapons will necessarily have to consider how its own actions will affect its rival. By instituting an atomic weapons program, it may trigger a corresponding program. Depending upon the particulars of the situation, therefore, the uncertainty connected with a rival's willingness to react by "going nuclear" could have a dissuasive influence on national decision making.

This same condition has also been noted as a motive condition. Thus it has a "Jekyll-and-Hyde" aspect—as does a nuclear alliance. Here empirical analysis will be required to determine the nature of the effects of this variable.

Risk of Unauthorized Seizure

It has been suggested that a government that perceives the possibility that its nuclear weapons might be seized by antigovernment groups will be disinclined to pursue the nuclear option, thus admitting its inability to control certain aspects of the national military power base. Perhaps most important, the greatest risk exists during the early years of the nuclear weapons program, before a strong command and control network can be devised. This would apply especially to the period during which an initial operating capability was being developed.

The risk of unauthorized seizure is also present when there is a possibility of organized opposition by the country's own military (or segments of the military). Spain is one example where the cloud of a military coup looms over the civilian government. Of course military governments are frequently challenged by disenchanted segments within their own services. A command and control system that is secure against terrorist attack might not be able to withstand an assault from inside the military chain of command.

Accordingly, I posited that the leaders of a prospective nth country would perceive the risk of unauthorized seizure as credible if (1) the nation had experienced three coups or revolutions in the preceding ten years, with (2) the most recent having taken place within the past three years.[24] This choice of time and frequency of occurrence is based upon the expected six-year program period discussed in chapter 2, indicating that the country in question could not expect to complete its atomic weapons program without the possibility of antigovernment intervention.

Preemptive Intervention by a Major Power

The issue has been raised, in a hypothetical manner, whether the possibility of preemptive intervention by a major power might dissuade nations from acquiring nuclear weapons. Here the reference to preemptive intervention *does not* connote action by an adversary but rather connotes a concerted effort by one or more nuclear weapons countries to enforce a nonproliferation norm through active policing and through force. To be sure, this notion is a long way from the IAEA safeguards system we have grown accustomed to. Yet from a historical perspective this has indeed been the prevailing situation in the Federal Republic of Germany and Eastern Europe since 1945 (Modelski 1959). Whether such a situation could evolve in other locations is a speculative question. Nevertheless, this condition clearly applies to the Eastern European nations and therefore deserves inclusion. Therefore, I coded this predictor variable as operative for Eastern European nations and considered it inoperative elsewhere.

Of course we must recognize that prospective proliferants may all face the possibility of a preemptive strike, either nuclear or conventional, by any "concerned" rival. The Israeli attack on an Iraqi research reactor in June 1981 is a graphic illustration. If indeed the fear of a preemptive *nuclear* strike by an adversary has significant dissuasive effects, then the motive factor nuclear adversary (discussed previously) should demonstrate little or no motive effect at all. Thus examining the nuclear adversary variable should reveal whether the incentive aspect or the disincentive aspect of having a nuclear adversary is dominant.

The same holds for the fear of a preemptive conventional strike by an adversary. Consider that a preemptive conventional strike is most likely to be expected from an adversary with overwhelming conventional military power. Therefore, if the disincentive aspect of confronting an overwhelming conventional military threat dominates the incentive aspect, no significant motive effect should be observed.

Thus we allow for three distinct preemptive threats to affect proliferation decision making in prospective nth countries.

A "Peaceful Reputation"

Finally, a number of researchers have suggested that several nations—Sweden, Switzerland, and India—have gone to considerable lengths to cultivate peaceful reputations. This variable is not necessarily synonymous with being declared neutral. The former refers to a general international image based on past behavior, but the latter merely signifies present status. Thus Austria, today is a declared neutral though it lacks a substantial history of peaceful behavior.

The acquisition of nuclear weapons, it is argued, would undermine the behavioral basis of the peaceful reputation. Therefore these countries' peaceful reputations and their corresponding international images act to dissuade them from "going nuclear." Since the nuclear proliferation literature is quite specific about who these nations are, peaceful reputations were attributed to Sweden, Switzerland, and India.

SUMMARY

This examination of the literature has produced a list of fifteen distinct motivational conditions, presented in table 8. If the motivational hypothesis is correct, these conditions should be systematically related to national decisions to initiate, or forgo, the manufacture of nuclear weapons. Indicators for each have been devised so we can determine their presence or absence each year at the nation level. Together there are 32,768 possible combinations of these fifteen variables, representing an equal number of possible motivational profiles. However, as I pointed out in the discussion, these fifteen variables reflect an even larger set of putative influences. And so an even larger number of possible constellations of political and military variables exists in the data base.

The indicators themselves were recorded simply as "present" or "absent," reduced to binary form. Although this measurement scheme may seem crude (wouldn't it be preferable to measure each variable along some kind of graduated scale?), it is probably consistent with the limitations and precision of the data and the models themselves. A higher level of scaling—or quantitative measurement—might introduce more noise than signal. Even in binary form, measurement error most certainly exists in the data. I assume, however, that the error is random (i.e., exclusion errors occur as often as inclusion errors) and that on an *average* the indicators do provide a fair representation of history.

But what exactly does a country's motivational profile look like, and how does it reflect complex situations in which several motive and dissuasive conditions are present at the same time? Consider India's situation in 1972 as defined by the variables described in this chapter. In 1972 India faced a nuclear-armed adversary (China, and perhaps from its perspective the United States); it had both regional and global power pretensions; it had acquired a nuclear ally (the Soviet Union), and India had a peaceful reputation. Letting zero represent the absence of the motivational condition and one represent its presence, we can represent the entire motivational situation described above as (100110000010001), where the position of each variable in the motivational profile corresponds to its variable number in table 8. As is apparent, the motivational profile is capable of reflecting any complex combination of motive and dissuasive conditions. Thus, by selecting any particular country at any particular point and recording the presence or absence of each of the fifteen conditions, we can obtain a proliferation-related motivational (or situational) profile of that country for that specific time.

TABLE 8
Summary of the Predictor Variables

Number	Name	Type
1	Nuclear threat	Both?
2	Latent capacity threat	Both?
3	Overwhelming conventional threat	Both?
4	Regional power status/pretensions	Motive
5	Global power status/pretensions	Motive
6	Pariah status	Motive
7	Domestic turmoil	Motive
8	Loss of a war	Motive
9	Regional nuclear proliferation	Motive
10	Defense expenditure burden	Motive
11	Nuclear ally	Both?
12	Legal treaties in force	Dissuasive
13	Risk of unauthorized seizure	Dissuasive
14	Possible nuclear intervention	Dissuasive
15	Peaceful reputation	Dissuasive

The real value of these motivational profiles lies in the capability to construct concisely a time-series history for each country. In a sense each motivational profile (corresponding to one nation-year pairing) represents a single focused-comparison case, and so these time series open up several possibilities for analysis. First, explicit tests of the motivational and sui generis hypotheses become possible. In the former case we should observe a systematic relation between specific forms of motivational profiles and proliferation decisions. In the latter case we expect a random relation. In both instances the time series prevents us from commiting an error common in case study research: selective focus on certain time periods or special cases. Here all countries and all years are equally weighted, and we are forced to account for cases where no proliferation decisions were made, for cases where they were, and for the timing of both.

Second, the time series can be used to investigate changes in the constellation of proliferation incentives and disincentives confronting individual countries over time. This kind of analysis can reveal some interesting aspects of the proliferation process as seen from the national perspective.

Finally, the motivational profile data could become a valuable basis for forecasting proliferation decisions, should the motivational hypothesis receive support from the data. Specifically, motivational profiles of countries of interest could be devised (based on projections and subjective assessments) for future years. The proliferation implications of those forecast motivational profiles could then be examined.

4

Is There a Technological Imperative?

Earlier I outlined three variations of the technological imperative hypothesis. While all three argue that possessing the technological capability to produce nuclear weapons is a sufficient condition for a decision to do so, each makes different assumptions about the nature of the relation between technology and decision. Since each variation also leads one to expect different forms of proliferation behavior, it should be possible to determine empirically which of the three models of the technological imperative most accurately portrays the nuclear proliferation process—if any of them do.

TECHNOLOGICAL IMPERATIVE I (MODEL I)

Model I is the most deterministic of the three technological imperative variants. Decisions to transform latent capacities into operational capabilities should quickly follow the acquisition of a latent capacity. Exactly what is meant by "quickly" is not stated, but it does not seem unreasonable to expect that a government will recognize its latent capacity and make its decision to push ahead with nuclear weapons production within several years of first acquiring a latent capacity.

Do proliferation decisions follow rapidly on the heels of acquiring latent capacities? Suppose we examine each nation and count the number of years it possessed a latent capacity. We can call this *latent capacity longevity*. For countries that eventually make proliferation decisions, latent capacity longevity is the number of years between the date of first acquiring a latent capacity and the date of the subsequent proliferation decision. The latent capacity longevity of the United States is zero years; for India (1965) it would be seven years. For countries that have not yet made proliferation decisions, latent capacity longevity is simply the number of years they have possessed a latent capacity to date. If model I is correct, most countries, if not all, should have short latent capacity longevities. There should be very few that have held latent capacities longer than several years. Moreover, if there really is this strong a systematic relation between having the technological capability to manufacture nuclear weapons and proliferation decisions, then the distributions of latent capacity longevity among proliferation decision cases and the remaining cases should be nearly identical. That is to say, the proportion of countries in the "proliferation decision" group that had latent capacity longevities of, say, zero to two years should be the same as the corresponding proportion in the "no proliferation decision" group. The reason for this is simple. The technological imperative hypothesis argues that all states with the capability to manufacture nuclear weapons will do so. Therefore countries that have not yet made proliferation decisions are "incubating." It follows that the distributions of the two groups must be the same (heavily bunched at the short end of the latent capacity longevity scale), so when these countries do make proliferation decisions and cross over to the proliferation decision group the latent capacity longevity distribution will be unaffected.

Table 9 presents some interesting findings, comparing latent capacity longevity of cases that involve proliferation decisions and of those that do not. The first column is the longevity period in

TABLE 9

COMPARISON OF "LATENT CAPACITY LONGEVITY

Latent Capacity Longevity (Years)	Cases in Which Proliferation Decisions Were Made	Cases in Which No Proliferation Decisions Were Made[a]
0-2	10 (77%)	2 (8%)
3-5	0	0
6-10	3 (23%)	3 (12%)
11-15	0	5 (19%)
16-20	0	6 (23%)
20	0	10 (38%)

[a]As of 1981 39 of the countries had latent capacities. The reader will notice that table 2 lists thirty-six countries. The three additional cases reflect restart situations for Britain (1947), India (1966), and South Korea (1975). In these three instances earlier proliferation decisions were made but aborted. Therefore their counting clocks (for latent capacity longevity) were restarted.

76

years. The second column is the distribution of proliferation decision cases; the third column is the distribution of the remaining cases. It is quickly apparent that nations that have made proliferation decisions have done so within a very short time of acquiring latent capacities. Roughly 77 percent of the proliferation decisions occurred within two years of acquiring latent capacities. In contrast, only 8 percent of those cases not involving proliferation decisions have latent capacity longevities under two years, while 80 percent of these cases have possessed latent capacities for more than ten years.

We see that countries that make proliferation decisions do tend to exhibit the kind of behavior expected by model I. There is a strong bunching of cases that did not hold latent capacities for more than two years before making proliferation decisions. They acquire the basic capability to manufacture nuclear weapons, then proceed to do so—just as the technological imperative hypothesis suggests. However, those cases where proliferation decisions have not been made exhibit behavior that is directly the reverse of what model I leads us to expect. The smallest proportion of cases have short latent capacity longevities, while the overwhelming majority have very long ones. It seems, then, that these two groups of countries exhibit opposite forms of behavior.

How likely is it that this apparent difference in behavior is merely due to random factors? How likely is it that no real difference exists between countries that decide to "go nuclear" and those that do not? A simple statistical test of the data in table 9 asks: Given the small sample sizes of the two groups, how likely is it that the apparent differences in their distributions are due to random fluctuations? What is the likelihood that, when the sizes of the samples are taken into account, there really is no true difference in their distributions? The Kolmogorov-Smirnov test is a very powerful nonparametric test—unlike many statistical techniques, it requires no assumptions about the distributional form of the data. It compares the cumulative frequencies of the two distributions. The test finds that the chance that there are no true differences between these two groups is less than one in 100,000. Fewer than one sample comparison in 100,000 would show a random difference as large as the one we observe here. Thus we may conclude that there is a very significant difference—both substantive and statistical—in latent capacity longevities between these two groups. This is not what model I of the technological imperative leads us to expect.

An alternative formulation of the expectations of model I— proliferation decisions should systematically follow latent capacities—is that the number of decisions to "go nuclear" in any given year should approximate the average number of countries acquiring a latent capacity during the previous several years. For example, if the number of nations acquiring latent capacities in four consecutive years was zero, one, two, and one, then on average we would expect to see one decision by the fourth year. This is a four-year moving average process in which the number of proliferation decisions in any given year is a function of the number of nations acquiring latent capacities in the several previous years. It assumes that the minimum decision delay is zero years, the maximum is three years and the average is 1.5 years. (In fact, the average delay for the thirteen cases of proliferation decisions is 1.7 years, while the median is zero.)

In figure 5, I have plotted the cumulative number of observed proliferation decisions and the expected cumulative number of

Fig. 5. The technological imperative (model I): A comparison of expected and observed proliferation decisions.

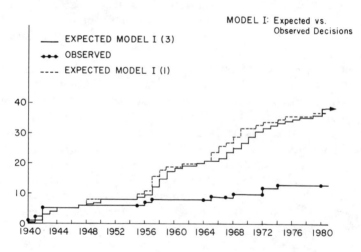

proliferation decisions for each year. The curve labeled "expected" represents the cumulative number of proliferation decisions we would expect if, on average, decisions systematically followed within several years the acquisition of a latent capacity. In essence, this moving average process with a three-year lag "allows" time for the influence of technology to gain momentum within the decision-making setting. It also takes into account decision-making inertia—the inevitable delays owing to uncertainty, procrastination, and risk-averse behavior on the part of decision makers. According to this formulation of model I, there should have been about thirty-eight proliferation decisions by 1981—almost three times as many as observed. The degree of divergence between the observed and expected curves is so great that any notion of a fit can be rejected.

What figure 5 does show, however, is that, through 1956, model I fits the observed pattern rather well. That is to say, from 1940 to 1956, government decisions to pursue nuclear weapons production were forthcoming as soon as indigenous capabilities allowed it. The cases I am talking about here are Germany (1940), Japan (1941), the United States and Britain (1942), the Soviet Union (1942), Britain (1947), and France (1956). With the possible exception of France, we know that each of these countries launched into nuclear weapons production at the earliest feasible moment—mostly within a year of acquiring a latent capacity. In this period, then, the technological imperative does indeed correctly "predict" proliferation decisions. Yet after 1956 the observed and expected curves diverge substantially. Whereas from 1957 to 1981 there was on average one proliferation decision every four to five years, model I expects an average of at least one proliferation decision every year.

A first reaction to these split results might be to speculate that a fundamental structural shift in the process of nuclear proliferation occurred about 1955-56. That is to say, before 1955-56 the process of nuclear proliferation was indeed driven by a technological imperative, but sometime in the mid-1950s the causal link between acquiring technological capability (latent capacity) and proliferation decisions was broken. Could, for example, the gratuitous giveaway of nuclear technology that characterized the Atoms for Peace program have confounded the relationship between technology and proliferation decisions?

When table 9 is split into two periods the result is table 10. The distributions for the period 1940-56 are precisely what we would expect to see under model I: few latent capacities exist for more than several years without subsequent proliferation decisions. For both groups—cases involving proliferation decisions and cases where such decisions have yet to occur—about 80 percent of their respective latent capacity longevity distributions lie in the category of zero to two years. Means and medians for the two groups show fair correspondence: mean longevities between one to two years and medians of zero. However, in the period 1957-81 the expected pattern breaks down. While the proliferation decision group appears to bunch at the low end of the distribution, the cases not involving proliferation decisions clearly spread out in the opposite direction. The latter group bunches at the long end of the latent capacity

TABLE 10
COMPARISON OF LATENT CAPACITY LONGEVITY

Latent Capacity Longevity (years)	1940-56		1957-81	
	Cases Involving Proliferation Decisions	Cases Not Involving Proliferation Decisions	Cases Involving Proliferation Decisions	Cases Not Involving Proliferation Decisions
0-2	6 (86%)	3 (75%)	4 (67%)	2 (9%)
3-5	0	0	0	0
6-10	1 (14%)	1 (25%)	2 (33%)	3 (14%)
11-15	0	0	0	5 (23%)
16-20	0	0	0	6 (27%)
20	0	0	0	6 (27%)
Average (years)	1.1	2.5	2.2	14.1
Median (years)	0	0	0	16

longevity scale. Model I behavior is not demonstrated in this later period. For cases involving proliferation decisions, the mean and median latent capacity longevities approximate those found in the earlier period (within statistical limits); here the earlier pattern persists. This is clearly not true for the cases not involving proliferation decisions. On average, during the period 1957-81, countries that have not opted to "go nuclear" have held latent capacities for fourteen to sixteen years. Thus, besides arguing against model I, the data also offer evidence of a fundamental change in the nuclear proliferation process in 1955-56. While model I of the technological hypothesis imperative may account for proliferation decisions in the earlier period, it does not account for those in the latter period.

Before accepting the implicit notion that more than one nuclear proliferation process has been at work, let us continue to test the remaining models in the hope that a single more parsimonious explanation for the *apparent* difference between these two periods will emerge. For instance, the same split pattern of behavior is possible under the motivational hypothesis if some of the key political variables occurred with unequal frequency in the two periods. And, as I describe below, a long-term version of the technological imperative also generates comparable expectations. For the moment, then, we put model I of the technological imperative aside because, at a minimum, it does not account for the most important segment of the data: contemporary events.

TECHNOLOGICAL IMPERATIVE II (MODEL II)

Technological opportunity may indeed be irresistible. But that does not mean that a technological imperative must operate in a stimulus/response fashion. In contrast to the short-run determinism of model I, model II emphasizes long-run determinism. Inevitably, all countries capable of producing nuclear weapons will do so. The exact timing of their decisions will depend on many factors, but proliferation decisions are inevitable.

Though this formulation of the technological imperative may not be as elegant as model I, it nonetheless suggests a certain pattern of expectations. First, if proliferation decisions systematically follow the acquisition of latent capacities but are randomly distributed in time, then we might expect the ratio of cumulative proliferation decisions to cumulative latent capacities to oscillate around a fairly stable average value. In other words, a relatively constant fraction of countries capable of manufacturing nuclear weapons should make decisions to do so over time. A high stable fraction (say above 80 percent) would actually be consistent with model I. A lower constant ratio would be consistent with model II, implying that the rate of proliferation decisions lags behind the rate of new latent capacities. A pattern that shows changes between two or more stable levels would suggest structural changes in the proliferation process itself, as was observed in the examination of model I.

Indeed, figure 6 suggests just such a pattern. Whereas two rather high plateaus can be seen from 1940 to 1954, a significant drop occurs from 1954 to 1964 and then stabilizes between 0.3 and 0.4

Fig. 6. The technological imperative (model II): The ratio of proliferation decisions to latent capacity opportunity.

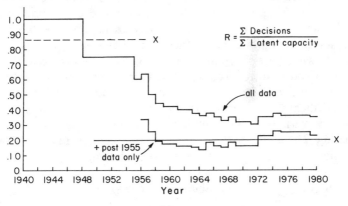

through 1980. The earliest period (1940-54) is consistent with model I, while the latest period (1964-80) is not. In one respect all the data, including the transitional drop (1954-64), may be considered consistent with model II. Given that the nuclear proliferation process first got under way in 1940, it would take time for the long-term technological imperative effect to stabilize at its "natural" level—at least this is what model II argues. And, indeed, if you look at the second curve in figure 6, representing the ratio of proliferation decisions to latent capacities starting in 1956, there is evidence to support this. It appears that a fairly stable proportion of the countries acquiring latent capacities after 1956 did in fact choose to convert them into operational capabilities. Overall, then, from the perspective of model II, these results suggest that the long-run rate at which nations are acquiring latent capacities is outrunning the rate at which proliferation decisions are being made by about four to one. Thus, for every five countries that acquire a latent capacity in the next decade one country will opt to go nuclear. If most of the world's industrial and industrializing nations (say about one hundred) acquire latent capacities by the year 2000, then this pattern of a technological imperative will give rise to a total of about twelve new proliferation decisions.

While the data are consistent with these initial expectations of model II, there are notable problems. First, the observed pattern in the data is also potentially consistent with several alternative hypotheses—including the motivational hypothesis and the null hypothesis. From the perspective of the motivational hypothesis, we would expect to see the post-1956 pattern of a fairly stable fraction of proliferation decisions if the appearances or manifestations of the relevant political and military variables themselves were evenly spread out in time and space. If this was true, then very strong mixes of these motivational variables would coalesce at a fairly stable rate. Correspondingly, this would cause proliferation decisions to occur with the stable frequencies observed above. In the sui generis world of the null hypothesis, this pattern is also exactly what we would expect to see—given the idiosyncratic nature of the hypothesized process. Where the motivational hypothesis would have proliferation decisions occur randomly in time but systematically with respect to the presence of certain political and military variables, the null hypothesis argues against the latter relation. The null hypothesis expects no systematic relation between proliferation decisions and the motivational variables.

A second problem is that, while the data do allow us to reject model I of the technological imperative in favor of model II, they do not allow us to accept model II as an explanation for proliferation decisions. Our initial test of model II only demonstrates that a fairly stable fraction of countries capable of producing nuclear weapons have decided to do so. But model II does not help us determine which countries will make proliferation decisions. Though I noted above that for every five countries that newly acquire latent capacities one will make a proliferation decision, this does not necessarily mean that any of those five will do so. Model II only expects that one new proliferation decision will arise. In other words, model II fails to link latent capacities to subsequent proliferation decisions other than to state that the former is *necessary* for the latter. For the technological imperative hypothesis to be a valid explanatory model, it must offer technological basis for discriminating the 20 percent that decide to "go nuclear" from the 80 percent that do not. If it cannot provide a systematic explanation, then, it cannot be distinguished from the null hypothesis. A third, related, problem is raised by the previous analysis of model I. I showed that, historically, once a nation had possessed a latent capacity for ten years the likelihood of its making a decision to "go nuclear" dropped to practically zero. Yet model II would expect nuclear decisions to be forthcoming at any point. The distribution of latent capacity longevity for the proliferation decision cases should therefore be spread out in a fairly flat manner. Yet we know that it is not.

This anomaly can be remedied if we restrict model II to argue that about 20 percent of all countries that newly acquire latent capacities will make proliferation decisions. This is consistent with my analyses. But, again, 20 percent does not reflect much of a technological imperative. Why doesn't technological momentum affect the other 80 percent of the countries? Or, alternatively, what is it about those 20 percent that make them so sensitive to the technological imperative? To try to answer these questions, let us examine a third version of the technological imperative hypothesis, one that does suggest an explanation: relative levels of nuclear infrastructure development.

TECHNOLOGICAL IMPERATIVE III (MODEL III)

Model III ties the rate of proliferation decisions to levels of nuclear development. "The more nuclear infrastructure a nation possesses, the more likely it is to move ahead to produce nuclear weaponry." It is important not to confuse model III of the technological imperative with the notion that, given a specific set of politico-military incentives, a higher level of nuclear development will make a proliferation decision more likely. The two models are not the same. In the former, technology causes proliferation decisions, irrespective of prevailing politico-military circumstances. In the latter, technology acts as an aid, facilitating decision making; but the requisite politico-military variables must be present.

Under model III, the more developed a country's level of nuclear infrastructure, the more likely it is to decide to "go nuclear." The strength of the technological imperative grows in direct relation to the size of the nuclear establishment in any given country. As more and more nuclear infrastructure is built up and the task of manufacturing nuclear weapons becomes easier and less formidable, the logic of the process pushes a government into a proliferation decision.

In figure 7 I have plotted the relative number of nations with latent capacities constituting (1) no nuclear infrastructure, (2) a moderate

Fig. 7. Prevailing nuclear infrastructures.

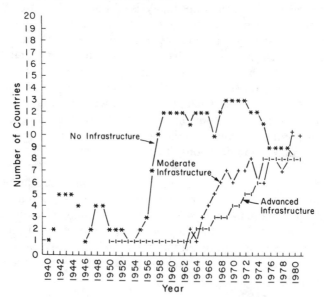

nuclear infrastructure, and (3) advanced nuclear infrastructure in each successive year. These correspond to countries that would require six years, four years, and less than two years to produce their first nuclear weapons. Notice that it is about 1955—the point at which Atoms for Peace got into full swing—that the number of nations with latent capacities really began to increase. Until 1963 or so, these countries all were lacking in significant nuclear infrastructures, and any nuclear weapons program would have required the full range of activities described in chapter 2. As of 1963 some countries began to acquire moderate nuclear infrastructures. Moderate nuclear infrastructures provided them with the means to produce fissile materials, thereby eliminating one crucial component from the resource demands of a future nuclear weapons program. The rate of growth of moderate infrastructures closely paralleled the initial growth rate of no infrastructure latent capacities in the previous period, though at a more modest pace. The lag between the two growth trends was of the order of eight to ten years.

Though the advent of advanced nuclear infrastructures begins about the same time as that of moderate nuclear infrastructures, the growth rate of the former is slower. With the jump to advanced infrastructures, prospective proliferants practically have a "free ride" in terms of nuclear weapons production. Note that the development of advanced nuclear infrastructures has lagged behind moderate infrastructures by just several years. In fact, when we examine the transition times for individual nations to move from a no infrastructure latent capacity to a moderate infrastructure latent capacity and finally to an advanced infrastructure, we find the following: transition from no infrastructure to moderate infrastructure averages 8.3 years (the median is 7 years). Transition from moderate infrastructure to advanced infrastructure averages 3.2 years (with a median of 1 year!). Thus the second transition takes less than half the time of the first transition. This certainly suggests technological momentum, at least in terms of the dynamics of nuclear infrastructure development.

When the data in figure 7 are summed, we find that for the period 1940 to 1981 there were 319 nation-years of no infrastructure latent capacities (58 percent), 121 nation-years of moderate infrastructure latent capacities (22 percent), and 109 nation-years of advanced infrastructure latent capacities (20 percent).[1] If Model III is correct we would expect to find the highest proportion of proliferation decisions among the cases involving advanced nuclear infrastructures and the lowest proportion among the cases with no nuclear infrastructure. In fact forty-eight (79 percent) of the sixty-one nation-year cases of proliferation decisions involved nations without nuclear infrastructures, six (10 percent) of the cases involved a moderate infrastructure, and seven (11 percent) involved advanced infrastructures. Thus model III does not seem to be supported by the data. We should have observed just the opposite distribution. Most of the cases of proliferation decisions should have come from the advanced infrastructure category.

In examining model I and model II, we found considerable evidence of a structural shift in the nuclear proliferation process. The shift in the linkage between technology and proliferation decisions is first noticeable about 1955-56. Therefore it might be revealing to examine model III using only the post-1956 data. This eliminates any confounding influences of the early (1940-45) scientific feasibility programs and concentrates on the period of significant growth in nuclear infrastructures. Between 1957 and 1981 there were 494 nation-years of latent capacities, of which 272 nation-years were of no infrastructure experience (55 percent), 121 nation-years of moderate infrastructure experience (24 percent), and 102 nation-years of advanced infrastructure experiences (21 percent). In comparison, of the twenty-eight nation-years among the proliferation-decision cases during this period, fifteen involved no infrastructures (54 percent), six involved moderate infrastructures (21 percent), and seven involved advanced infrastructures (25 percent). Again this is the reverse of what model III leads us to expect.

In fact, for 1957 to the present the relation between the level of nuclear infrastructure and proliferation decisions appears to be random. When the distribution of infrastructure categories among the proliferation decision cases is compared with the corresponding distribution among cases involving no proliferation decisions, the two groups show no systematic differences. Two statistical tests are useful: the chi-square test and the Kolmogorov-Smirnov test. The chi-square test compares the relative frequencies of each infrastructure category among the proliferation decision cases against the same relative frequencies among the remaining (no proliferation decision) cases. Noting differences between observed and expected frequencies, it gives the likelihood that there are no systematic differences. The result here is about an 85 percent chance that no systematic difference exists (*chi-square = 0.33 with 2 degrees of freedom).

The Kolmogorov-Smirnov test asks the same question, but it uses other characteristics of the data. First it takes into account the ordinal quality of the data—that is, that there is a natural ranking to the data from no infrastructure to advanced infrastructure categories. Second, the Kolmogorov-Smirnov test uses the cumulative frequencies rather than the relative frequencies of the data. The two tests can be used to "check" one another, since they use different information from the data. The Kolmogorov-Smirnov test gives a probability of 99 percent that there is no systematic difference. (The maximum difference in cumulative frequencies is 0.05.) Both tests agree that model III expectations are not met; proliferation decisions do not appear to be related to levels of nuclear infrastructure.

A somewhat different interpretation of model III is possible. Rather than examining the nuclear infrastructure level of the prospective proliferant, we could consider whether the technological momentum toward nuclear proliferation is generated at the system level. One could argue that the global move toward nuclear energy that occurred between 1960 and 1975 could also have triggered interest in nuclear weapons. In this way it would be the level of global nuclear infrastructure—not that of the proliferants themselves—that creates the impulse for proliferation decisions. Accordingly, as the *global* level of nuclear development increases, we should observe an increase in proliferation decisions. Here nuclear development connotes both the quantity and the quality of national latent capacities. From figure 7 we can see that the time span breaks clearly into three segments: from 1940 to 1955 few nations had latent capacities and none of these had nuclear infrastructures. From 1956 to 1965 the number of nations with latent capacities rose rapidly— but few nuclear infrastructures had yet developed. Finally, from 1966 to 1981 nuclear infrastructures grew at least as rapidly as the number of countries acquiring basic latent capacities without infrastructures. Model III behavior would predict the lowest rate of proliferation decisions in the earliest period, when few nations had latent capacities and none had significant nuclear infrastructures. A higher rate of proliferation decisions should be observed in the second period, when a large number of countries came to acquire latent capacities. And last, the highest rate of proliferation decisions should be seen in the third period, when moderate and advanced nuclear infrastructures came to dominate.

In fact, the corresponding rates are 0.38 (six decisions in sixteen years), 0.3 (three decisions in ten years), and 0.27 (four decisions in fifteen years). While it is statistically probable that these three rates are merely random deviations from an overall average rate of about 0.32, there is no way they could be manipulated to portray the kind of behavior model III predicts. In other words, the rate of proliferation decisions over time may be decreasing slightly or may be fairly stable, but it clearly is not increasing as model III of the technological imperative leads us to expect. A parallel, and perhaps more rigorous, test can be run by observing that since 1956 there has been a steady and monotonic increase in the global spread of nuclear technology. Leaving aside the specific measures of nuclear infrastructure levels used above, from 1956 on each subsequent year has seen the global stock of nuclear hardware grow to some degree— more power reactors, more fuel fabrication plants, more pilot scale facilities. Thus, if model III is correct, the time between successive proliferation decisions should be growing shorter as time goes on. We can test this by ranking the periods between successive proliferation decisions in terms of length and then computing rank correlation coefficients against rank in time. Interdecision times should shorten as time progresses. The result is a rank correlation of 0.14 but with a probability of significance of more than 0.7— meaning there is more than a 70 percent chance that the rank correlation is actually zero. In other words, there is no trend among the interdecision time periods, and model III is not supported.

SOME OBSERVATIONS

I have been unable to find consistent support for any of the models of the technological imperative commonly described in the nuclear proliferation literature. When the entire time series was examined, the inconsistencies between the data and expectations derived from the models were numerous and strong. One cannot distinguish between countries that decide to "go nuclear" and those that do not on the basis of nuclear technology. The opportunity to "go nuclear" is not systematically followed by proliferation decisions (models I and II). Countries with more highly advanced nuclear infrastructures do not seem to exhibit any greater nuclear propensity than countries lower on the nuclear development scale (model III). Empirically, there is no compelling case for the argument that nuclear technology causes nuclear proliferation. It seems then that we are on fairly secure ground in rejecting the technological imperative hypothesis as a single explanation for decisions to initiate nuclear weapons programs.

But what role does technology actually play in nuclear proliferation? Let us reexamine some of the findings of this chapter. The great majority of countries that "go nuclear" do so very quickly after acquiring latent capacities. None of the countries that made proliferation decisions had a latent capacity longevity of more than ten years. In contrast, 80 percent of the countries that have not made proliferation decisions have latent capacity longevities of ten years or more. From table 9, it is easy to compute that, in purely statistical terms, the likelihood of a country's deciding to "go nuclear" if it has had a latent capacity for ten years or less is about 33 percent. But that likelihood approaches zero after the country passes the ten-year point. Thus we might think in terms of a ten-year "proliferation window" that each country with a newly acquired latent capacity must pass through. If a country is going to decide to go nuclear, it will do so within the first ten years in which it has a latent capacity. Once a country passes through this window, it is very unlikely to attempt to "go nuclear."

This notion of a "proliferation window" is not meant to imply any kind of behavioral law. It is a description of the nuclear regime over the past thirty years or so. Since for nations with latent capacities the technical task of manufacturing nuclear weapons gets easier as time progresses, perhaps the result is a reverse technological imperative. As a nation's capacity to produce nuclear weapons increases—as the difficulty, burden, and time lag decrease—the actual need to make the bomb components may decrease. The stronger one's option, the less the momentum to invoke the option. Over time an institutional perspective may develop—a kind of decision inertia—in which only a major crisis would be sufficient to provoke a proliferation decision.

The analyses also offered some reason to believe that a structural shift in the nuclear proliferation process had occurred sometime between 1955 and 1958. In particular, by 1958 proliferation decisions seem to have become decoupled from the simple acquisition of latent capacities. Before 1956, possession of a latent capacity almost certainly indicated a strong likelihood of "going nuclear." After 1956 this relationship fell apart.

The period 1940-56 is, of course, the time when the world's major powers initiated their nuclear decisions: The United States, Britain, the Soviet Union, France, Nazi Germany, and Imperial Japan. It may be that the technological imperative seemed to fit during this period because high levels of technology and industry are strongly correlated with major power status. If a similar or interrelated set of variables endemic to major power status accounts for these major-power proliferation decisions, then, regardless of the true reason for these countries' pursuit of nuclear weaponry, we can expect a strong but spurious correlation between latent capacity and prolif-eration decisions. An equally high correlation would be found, for example, if we used battle tank production as an indicator of future proliferation decisions, because all the major powers had very high levels of tank production.

The period after 1956 demonstrates no systematic relation be-tween latent capacity and proliferation decisions. This latter period is characterized by non-European proliferation decisions. Here relative levels of technology and industry are much more varied, as are political, social, military, economic, and geographical variables. Significantly, it was during this period that the basis of acquiring a latent capacity switched from indigenous efforts to being grounded largely in foreign assistance. International programs that provided research reactors almost on demand greatly accelerated the rate at which latent capacities were acquired. The large-scale release and active dissemination of information pertaining to plutonium reprocessing (1958-65) further lowered the latent capacity threshold and reduced the resource demands of a nuclear weapons program. The result: many more nations acquired latent capacities, and many did so much sooner than otherwise would have been the case. Where in the earlier period those nations most capable of initiating a nuclear weapons program were all major powers, in the latter period political, economic, military, and social diversity characterized the pool of prospective proliferators.

Since 1957, about 20 percent of those countries that could have opted to "go nuclear" have done so. This rate has been remarkably stable over the past twenty-five years. The analyses of latent capacity longevity show clearly that this stable rate is *not* the product of queuing—that is, many countries with newly acquired latent capacities in the process of going nuclear.

Finally, I found that the transition times between the three nuclear infrastructure categories grow shorter as one moves up the scale. It takes less than half the time to move from moderate to advanced nuclear infrastructure as it does to move from no infrastructure to moderate infrastructure. This pattern would have continued strongly had it not been for United States initiatives to halt plutonium recycling in the late 1970s. United States interventions to halt the acquisition of commercial reprocessing facilities by South Korea and Pakistan are just two examples.

Technology is certainly a necessary condition for nuclear proliferation. The analyses presented in this chapter demonstrate that it is not a sufficient condition. As it pertains to nuclear proliferation there is no technological imperative, but this does not mean that technology does not affect proliferation decision making. A country's relative level of nuclear development can act as a conducement to proliferation decisions. On the one hand, it seems intuitive to argue that as one's level of nuclear infrastructure advances it becomes *easier* to make proliferation decisions—given one is prone to do so. For example, the relative resource requirements of a nuclear weapons program could affect the politico-military calculus invoked by the motivational hypothesis. Or it might make the whim of a mercurial leader that much easier to fulfill—as is suggested by the null hypothesis. Of course the data examined in this chapter contradict this line of reasoning. In fact they argue for just the opposite: as the resource demands decline, proliferation decisions are less likely. The propensity to put off the proliferation decision may increase as the costs of doing so decline.

The question still remains: Why do those countries that opt to "go nuclear" do so?

5
Testing the Motivational Hypothesis

While the motivational hypothesis offers a relatively simple explanation for proliferation decisions—politico-military influences—it also calls into play a fairly complicated model of decision making. Individual motive conditions may act independently or collectively in moving a nation toward nuclear weapons production. National decision making, however, may or may not be influenced, depending on the presence of other variables—dissuasive conditions. Indeed, the fifteen motivational variables described previously offer 32,768 possible combinations—almost 33,000 distinct motivational profiles are possible. In contrast, the technological imperative hypothesis invoked but one major determinant: relative levels of nuclear development.

In parallel to the approach followed in testing the technological imperative hypothesis, the task here is to establish expectations based on the motivational hypothesis and then compare them with the data. Which combinations of motivational variables should be systematically associated with proliferation decisions, and which combinations should not? Can we correctly pick those historical cases where proliferation decisions were made, using only knowledge of the prevailing motivation profiles, without erroneously selecting cases that did not involve proliferation decisions? This is, after all, the test the technological imperative hypothesis failed.

To set up expectations based on the motivational hypothesis, some preliminary developmental work is necessary. First, the concept of *nuclear propensity* is introduced—the extent of a nation's explicit (but time-varying) predisposition toward inititiating the manufacture of nuclear weapons. In the context of the motivational hypothesis, a nation's nuclear propensity is the degree to which national decision makers are confronted with many incentives (and few significant disincentives) to "go nuclear." Nations with "strong" nuclear propensities are expected systematically to make proliferation decisions. Those with "weak" nuclear propensities are expected to possess ever-present latent capacities but not to make any direct decision to develop operational capabilities.

I have explicitly avoided using the term "probability." I did so to short-circuit temptations to make inferences that normally would be appropriate where the rules of classical probability theory apply, but that would be inappropriate in the context of this analysis. In particular, it is not at all clear that many of the well-defined properties of probabilities delineated by classical probability theory would be applicable to the phenomena being examined here. Multiple trials and replication of events are two examples. Suppose Switzerland was "found" to have a consistent nuclear propensity of 10 percent. Interpreting 10 percent as a probability would imply that over a twenty-five-year period the cumulative probability that Switzerland would decide to start a nuclear weapons program would surpass 90 percent. However, viewed as a measure of propensity, where zero reflects no nuclear propensity and 100 percent reflects an absolute "compulsion," then 10 percent would signify a relatively low nuclear propensity. Therefore a 10 percent nuclear propensity over a twenty-five-year period would suggest no significant inclination toward wanting to manufacture nuclear weapons. In this respect this concept of nuclear propensity is most similar to the Bayesian notion of subjective probability—statements about the strength of belief (certainty) that some event will occur. A 10 percent nuclear propensity and a 90 percent nuclear propensity both reflect the same degrees of certainty, though at opposite ends of the likelihood spectrum. The former suggests a high degree of certainty that no proliferation decision is forthcoming; the latter implies an equal degree of certainty that a proliferation decision is imminent. In contrast, a 50 percent nuclear propensity reflects the greatest uncertainty—in a sense, even odds. A proliferation decision may or may not be forthcoming. So, while the computational procedures discussed below are those commonly used in probability analysis, it is important to keep in mind that their interpretation is quite different.[1]

In the way of generating expectations, then, the motivational hypothesis postulates that weak nuclear propensities do not lead to proliferation decisions, strong nuclear propensities are systematically associated with proliferation decisions, and moderate nuclear propensities result in vacillation, ambiguity, and mixed forms of proliferation-related behavior. Nations with weak and moderate nuclear propensities may engage in many activities designed to keep their nuclear options open (perhaps the latter group even more than the former), but they are not expected to make decisions to transform their "options" into operational capabilities.

What remains is to derive nuclear propensities from the motivational profiles. Is it possible to determine a country's nuclear propensity at a given time by examining the prevailing constellation of political and military variables? The basis for doing this is found in the nuclear proliferation literature itself—specifically, in the ways incentives and disincentives have been postulated to influence proliferation decision making. We are interested in ways the motivational hypothesis—or, more appropriately, those who subscribe to it—portray interactions between and among the motivational variables and how these interactions in turn are believed to affect nuclear propensities.

First, note that the motivational hypothesis views proliferation decisions as the outcome of positive political decision making. That is to say, decisions to "go nuclear" are forthcoming when there are specific reasons to do so (incentives outweigh disincentives) rather than because of a lack of disincentives. In this respect the motivational hypothesis argues that there is no *ambient* global tendency toward nuclear proliferation in the absence of politico-military incentives.[2] It follows, then, that motivational profiles lacking any motive conditions as components should correspond to low nuclear propensities—actually, zero nuclear propensities.

Second, discussions in the literature imply that each individual motive condition should demonstrate a detectable positive impact on nuclear propensities. Here the implication is that none of the motive conditions acts as a mere catalyst. Rather, each should possess an intrinsic influence. Of course the relative effects of the motive conditions may be quite different, and there may indeed be interactive effects in combination. However, the incentives connected with any given motive condition are not altered by the presence of any other motive conditions. So again independent influences are preserved.

A third assertion pertains to the characteristics of the dissuasive conditions and how they influence nuclear propensities. In particular, the extent to which a specific dissuasive condition can reduce nuclear propensities will depend upon the array of motive conditions that constitute the associated motivational profile. In other words, individual dissuasive conditions affect nuclear propensities indirectly by weakening the relative influence of the individual motive conditions.[3] Some dissuasive conditions may exhibit no weakening influence on certain motive conditions. Being a signatory to the NPT may effectively counter the status incentives a regional power may perceive in "going nuclear," but it may offer no counter to the incentives sensed by a pariah confronting an overwhelming conventional military threat. Thus, in contrast to the assumption of how the motive conditions operate, the effect of dissuasive conditions is explicitly tied to the presence of specific motive conditions.

Fourth, the association between the relative presence of the motivational variables and a proliferation decision need only be probabalistic. The motivational hypothesis does not argue that any form of determinism exists in proliferation behavior. In much the same way that one is more likely to forecast rain when it is cloudy than when it is clear, so too might one expect proliferation decisions when "conditions" are favorable. Of course, sometimes it will not rain—and sometimes proliferation decisions may not occur when expected. It is important not to confuse this notion of probabilistic relationship with randomness. The motivational hypothesis expects systematic relationships but recognizes that perfect associations are unlikely.

One final assumption that plays an important role is decision reversibility. Simply stated, up to the point at which the first nuclear weapons are produced, proliferation decisions can be reversed. The transformation of a latent capacity into an operational capability can be stopped before weapons fabrication—albeit with a likely significant increase in the potential inherent in the latent capacity. India's 1966 reversal of the previous year's decision to press ahead with a nuclear explosives test and South Korea's decision to cancel its nuclear weapons project in 1975 are illustrations. Thus, if the motivational hypothesis is correct, we should observe high nuclear propensities throughout the time interval between the initial proliferation decision and the production of the first nuclear weapon. Conversely, should nuclear propensities drop out of the strong region, we should see a decision reversal. (This sets up a much more stringent test of the motivational hypothesis than does accounting "simply" for initial proliferation decisions.) Of course "decisions" not to manufacture nuclear weapons can also be reversed.

A First Test

The propositions outlined above set up a number of expectations that can be used for an initial test of the motivational hypothesis. In its simplest form, the motivational hypothesis leads us to expect a proliferation decision whenever a motive condition is present alone or in combination with other motive conditions. (No dissuasive conditions, however, are present.) If this is historically true, then a quick test of the data should reveal it. Suppose we examine only those cases in which motive conditions are present and dissuasive conditions are absent. We should see a significantly larger proportion of cases involving proliferation decisions among this subset of cases than we see in our data set as a whole. If there is a systematic relation between proliferation decisions and motive conditions acting in the absence of dissuasive conditions, then many more proliferation decision cases should be among this subset of cases than we would expect by chance. For example, suppose we examined the subset of cases involving overwhelming conventional threats. If 11 percent of our entire data set corresponds to proliferation decisions, then random chance would see about 11 percent of this subset of cases involving proliferation decisions. Should there be significantly more than 11 percent, it would imply a systematic relation between the presence of this motive condition and proliferation decisions and hence support the motivational hypothesis. Should there be significantly fewer than 11 percent, it would also imply a systematic relation, but one implying dissuasive effects.

Table 11 gives the results of this simple test. I tabulated the proportions of proliferation decisions observed among the subset of cases for each motive condition (i.e., the fraction of proliferation decisions given the ith motive condition was present), but again only in those instances in which dissuasive conditions were not present. These are found in the observed column. Then I calculated the proportion expected via random sampling. Expected proportions are simply the overall proportion of cases involving proliferation decisions: 11 percent. If the observed and expected columns agree, then the motivational hypothesis is not supported. The results of the difference of proportion test (p in the third column) gives the probability that the difference between the observed proportion and the expected proportion can be attributed to random chance—that no true systematic difference exists.

As is apparent, the results are consistent with the expectations of the motivational hypothesis. In all but a single instance, the proportion of proliferation decisions observed is substantially greater than we would expect if only random factors were at work. Eight of the eleven variables demonstrate almost no likelihood of a random relation to the data set as a whole. Only domestic turmoil shows a relation that cannot be considered representative of statistical agreement between observation and expectation. Thus there is some

TABLE II
FIRST TEST OF THE MOTIVATIONAL HYPOTHESIS: MOTIVE CONDITIONS

Motive Condition	Proportion of Proliferation Decisions Observed (%)	Proportion Expected[a] (%)	p[b]	N
Nuclear threat	100	11	0	15
Latent capacity threat	88	11	0	41
Overwhelming conventional threat	96	11	0	23
Regional power status/pretensions	93	11	0	42
Global power status/pretensions	97	11	0	37
Pariah status	100	11	0	9
Domestic turmoil	20	11	.32	10
Loss of a war	90	11	0	8
Regional nuclear proliferation	—	11	—	—
Defense expenditure burden	90	11	0	31
Nuclear ally[c]	4	11	.0004	200

[a]Rounded to nearest whole number.
[b]Probability that the difference between the observed and expected could be due to random chance. This is a one-tailed test because we have theoretical reasons to believe that the observed proportion will be greater than the observed portion. The cases are nation-year cases.
[c]This relationship is in the *reverse* direction expected; it cannot be a motive condition.

reason to suspect that domestic turmoil may lack the motive effects often attributed to it.

The four variables posited as potentially having both motive and dissuasive effects demonstrate opposite tendencies. Latent capacity threat clearly reveals motive condition influences and thus is considered as such for the rest of the study. Nuclear threat and overwhelming conventional threat both demonstrate motive condition tendencies, and therefore it seems we can rule out any significant effects from associated disincentives related to fears of a preemptive strike by an adversary. (See the discussion in chap. 3.) However, nuclear ally demonstrates an equally clear dissuasive effect. Significantly fewer proliferation decisions occur than we would expect in random sampling. Thus nuclear ally will be considered a dissuasive condition.

So our initial look at the motive variables suggests remarkable consistency with the expectations of the motivational hypothesis. Equally important to its validation, however, are the influences of the dissuasive conditions. Each dissuasive condition should exhibit a dampening influence on at least a couple of the motive conditions. In other words, the proportion of proliferation decisions among those cases that include both a specific motive condition and a relevant dissuasive condition should be significantly lower than what we find when only the motive condition is present. The results for this test are shown in table 12, where only the significance test probabilities are given.

Here too the motivational hypothesis is supported. Each of the proposed dissuasive conditions appears to be associated with a real reduction in the proportion of proliferation decisions associated with several motive conditions. Probability values below 0.01 can be used as a useful cutoff point in deciding statistical significance—though so many values are approximately zero that no such arbitrary decision is necessary. Thus I shall retain all the dissuasive conditions for the study.

Let me caution the reader not to try to assess any one-on-one

TABLE 12

FIRST TEST OF THE MOTIVATIONAL HYPOTHESIS: DISSUASIVE
CONDITIONS

Motive Conditions	Probability of No Real Reduction[a]				
	Nuclear Ally	Treaty	Unauth- orized Seizure	Preemptive Strike by Major Nuclear Power	Peaceful Reputation
Nuclear threat	0 (325)	0 (174)	.002 (14)	0 (139)	0 (57)
Latent nuclear threat	0 (193)	0 (103)	.08 (10)	0 (128)	0 (57)
Overwhelming conventional threat	0 (9)	0 (27)	.001 (10)	—[b]	0 (54)
Regional power status/ pretensions	0.(152)	0.(51)	0 (21)	0 (26)	0 (17)
Global power status/ pretensions	0 (60)	0 (28)	.33 (7)	0 (23)	.34 (3)
Pariah status	—	0 (6)	—	—	—
Domestic turmoil	.32 (22)	.17 (4)	.39 (12)	.24 (2)	.30 (10)
Loss of war	.008 (2)	—	—	—	.004 (6)
Regional nuclear proliferation	— (48)	— (16)	— (3)	— (14)	— (6)
Defense expenditure burden	.0 (9)	.003 (2)	.03 (8)	—	—

[a]N$_1$ corresponds to N in table 11. N$_2$ is given in parentheses here.
[b]Dash indicates no data.

relationships between motive and dissuasive conditions from table 12. While it is a good test of the motivational hypothesis as a whole, it addresses only one aspect of the postulated expectations. In particular, it considers only the statistical significance of differences, not their magnitude. Depending on the pertinent sample sizes, a marginal difference can have greater statistical significance than a much larger difference.

ASSESSING NUCLEAR PROPENSITIES FROM MOTIVATION PROFILES

Statistical significance is a much misused concept. It is a measure of relative independence—in our case, of how likely *apparent* differences between observation and expectation are to be true, systematic differences. In significance testing, probabilities close to zero imply very little likelihood that what we observe is merely a random variation of what we expected. The observed differences are real. However, a statistically significant relationship may be predictively (and analytically) insignificant. That is to say, the differences between observed and expected values may be real and systematic but may also be so small in magnitude as to be substantively unimportant.

Consider, for example, an instance where the proportion of proliferation decisions associated with regional proliferation is 70 percent. Suppose that when the dissuasive condition nuclear ally is also present the proportion of proliferation decisions observed drops to 65 percent. If the number of cases among the two samples rises above 150, a statistically significant difference will always be indicated. The drop in proportions from 70 percent to 65 percent would indeed indicate a systematic decline. (Of course smaller samples require larger differences to be statistically significant and therefore are more likely to reflect substantively significant differences.) Yet for our purposes the magnitude of the difference is almost negligible. According to the motivational hypothesis, the effect of the dissuasive conditions should be strong enough to produce changes in nuclear propensities large enough to alter the decision calculus of decision makers. It follows that a rigorous test of the motivational hypothesis must therefore examine the strength of the relation—in particular, predictive power—between the motivational variables and proliferation decisions.

The test of predictive power suggested above is to use the motivational profiles to predict proliferation decisions. Does knowledge of the motivational profiles yield substantial improvements in correctly guessing which nations make proliferation decisions, when compared with "blind" guessing? This is where the notion of nuclear propensities enters the picture.

The extent to which a specific motive condition is associated with proliferation decisions—and hence a measure of its contribution to nuclear propensities—can be gauged by the proportion of instances in which the motive condition is observed and is subsequently followed by a proliferation decision (say within a year). This tallying can be done in the presence of any other motive conditions but in the absence of any associated dissuasive conditions. Thus the simple nuclear propensity associated with a given motive condition is the number of cases in which that motive condition is followed directly by a proliferation decision, divided by the total number of cases in which that motive condition is present (when no dissuasive conditions are present). The result is an estimate of the conditional likelihood of initiating a nuclear weapons program given that a specific motive condition was first observed. So, if one hundred nation-year cases of adversaries with latent capacities are recorded and they are associated with forty *nation-year* instances of proliferation decisions, then the nuclear propensity associated with the motive condition adversary with a latent capacity would be 40 percent.

How do we estimate compound nuclear propensities when more than one motive condition is present? This is where the assumption regarding independent effects among the motive conditions is important, for it implies that we can use a simple multiplicative rule to determine compound nuclear propensities. Compound nuclear propensity is one minus the product of the simple nuclear propensities associated with the relevant motive conditions after each is subtracted from one. Thus two motive conditions with simple nuclear propensities of 60 percent and 80 percent would yield a compound nuclear propensity of 92 percent—that is, $1-(.4)(.2)=.92$. This is a simple but powerful formulation.[4]

We must also take into account the dampening effects of the dissuasive conditions. Just as we were able to calculate the conditional likelihood of a proliferation decision given a specific motive condition, we can estimate the revised nuclear propensity in conjunction with a dissuasive condition as the number of cases in which a given motive condition/dissuasive condition pair is followed directly by a proliferation decision, divided by the total number of instances in which the motive condition/dissuasive condition pair occurs. Again, this is a simple calculation of joint likelihood of occurrence.

One additional step is necessary to generate estimates for the simple and revised nuclear propensities of the motive conditions. Let us turn to some of the fundamental concepts of Bayesian statistics. Where classical statistics would simply use the estimates obtained from the likelihood calculations outlined above, Bayesian statistics argues that we have more useful information than just those observed frequencies. Specifically, if the motivational hypothesis is correct, then we have prior information that each motive condition is associated with a strong (simple) nuclear propensity. Likewise, each dissuasive condition should degrade some motive conditions to the point where the associated revised nuclear propensity is weak. If the motivational hypothesis is correct, this prior information—when combined with the observational data—should demonstrate improved predictive power. *If the motivational hypothesis is not valid, then the incorporation of the prior information will make the "lack of fit" that much more obvious.*

If we had no prior expectations regarding the effects of the various motivational variables, we would choose what is called a diffuse prior distribution that reflects the total uncertainty regarding nuclear propensities—all nuclear propensity values are equally likely. This prior information is factored into the calculation of simple and revised nuclear propensities by adding a 1 to the numerator and a 2 to the denominator. For example, if there were fifteen cases of proliferation decisions coupled to the motive condition nuclear threat and only fifteen cases involving nuclear threat (neither with any dissuasive condition present), then the simple nuclear propensity without the prior distribution would be 1.0. With the prior distribution the nuclear propensity is about 0.94. This particular prior distribution reduces the empirical estimate when it is greater than 0.5 and increases it when it is less than 0.5, since initially all values for nuclear propensity between 0 and 1 are viewed as equally likely—that is, a priori there is complete uncertainty. Note that the effects of any prior information decrease rapidly as the observed sample size increases, and so the diffuse prior distribution has a noticeable effect only when there is relatively little observational data present—just as we would like for predictive purposes and hypothesis testing.

The fact is that the motivational hypothesis does provide us with prior beliefs about the effects of the motivational variables. So the diffuse prior state is not appropriate here. Logical choices for the priors might be [2, 3] for simple nuclear propensities and [1, 3] for revised nuclear propensities. Adding 2 to the numerator and 3 to the denominator in calculating simple nuclear propensities and, correspondingly, 1 to the numerator and 3 to its denominator in calculating revised nuclear propensities has interesting intuitive appeal. First, these are the first integer priors directly above and below the diffuse prior state. They suggest greater certainty initially but will not overpower the observational data in moderate-sized samples. Second, they reflect *equal* degrees of certainty—though in opposite directions. Third, they reflect two-to-one odds, which establishes a wide and equally weighted discrimination range: a value from 0 to 0.33 connotes weak nuclear propensity; a value from 0.34 to 0.66 connotes a moderate nuclear propensity; and anything over 0.67 connotes a strong nuclear propensity. The results of this procedure are estimates of simple nuclear propensities and revised nuclear propensities that establish more stringent requirements for testing the motivational hypothesis.

Finally, when several motive conditions and dissuasive conditions are present we can estimate the aggregate nuclear propensity. This is done in several stages. First we calculate the simple nuclear propensity associated with each relevant motive condition. Next we calculate the revised nuclear propensity for each relevant motive condition, paired with each relevant dissuasive condition. Third, for each motive condition, we compute the adjusted nuclear propensity by multiplying each of its revised nuclear propensity values together successively (i.e., there will be one revised nuclear propensity value for each dissuasive condition present), then dividing this product by the motive condition's simple nuclear propensity raised to the $n-1$ power (where n is the number of dissuasive conditions present). This procedure is repeated for each motive condition. Finally, we calculate the aggregate nuclear propensity—which reflects the estimated nuclear propensity of the case at hand—by subtracting the adjusted nuclear propensity of each relevant motive condition from one, multiplying the results, and then subtracting the product from one. (These procedures are given in symbolic form in Appendix D.)

If this all seems confusing, it is because this simple series of likelihood models attempts to faithfully replicate the theoretical linkages described in the nuclear proliferation literature. Though the linkages are simple, they are woven together in a complex manner. In essence these formulations are merely restatements of what many have hypothesized as characterizing dynamics of the nuclear proliferation decision process.

Simple Nuclear Propensities and Revised Nuclear Propensities

Table 13 presents the simple nuclear propensities for each of the motive conditions and the corresponding revised nuclear propensi-

TABLE 13
NUCLEAR PROPENSITIES

Motive Conditions	No Dissuasive Conditions	Dissuasive Conditions				
		Nuclear Ally	Legal Treaty	Unauthorized Seizure	Preemptive Strike by Major Nuclear Power	Peaceful Reputation
Nuclear threat	.94	.02	.19	.52	.24	.08
Latent capacity threat	.86	.04	.17	.61	.14	.08
Overwhelming conventional threat	.92	.33	.09	.46	.33[a]	.03
Regional power status/pretensions	.91	.05	.29	.33	.54	.25
Global power status/pretensions	.95	.14	.21	.8	.25	.66
Pariah status	.91	.33[a]	.3	.33[a]	.33[a]	.33[a]
Domestic turmoil	.33	.16	.27	.26	.2	.3
Loss of a war	.81	.2	.33[a]	.33[a]	.33[a]	.22
Regional nuclear proliferation	.66	.07	.44	.33	.49	.22
Defense expenditure burden	.88	.16	.2	.54	.33[a]	.33[a]

Note: Estimates are based on 522 nation-year cases.

[a]Estimates based on Bayesian priors only; no empirical data.

ties in the presence of the various individual dissuasive conditions.

Looking down the first column, we see that most of the motive conditions have very strong associated nuclear propensities. Five of the motive conditions have associated nuclear propensities over 90 percent. These are nuclear threat, overwhelming conventional threat, regional power status/pretensions, global power status/pretensions, and pariah status. Instances where these motive conditions exist without any dissuasive factors will have high certainty of involving proliferation decisions. Still strong, but somewhat lower, nuclear propensities are associated with defense burden and loss of a war. Here proliferation decisions are still to be expected, but the certainty levels are reduced. Regional proliferation shows a nuclear propensity that is right at the margin of the moderate/strong regions. While still an important contributor to proliferation behavior, its effect is considerably less than that of the other motive conditions. In contrast, domestic turmoil shows only a moderate nuclear propensity influence (it is practically in the weak region). Once again the validity of including this variable in the model is called into question. Historically, domestic turmoil does not appear to have had a significant influence on proliferation decision making. Nonetheless, because of the role that has been ascribed to it in the future—particularly where Third World proliferation is expected to be the norm—it is worth retaining this hypothesized motive condition.

Scanning the other columns, we can now qualitatively assess the relative "effects" of the various dissuasive conditions in terms of their impact on the motive conditions. For example, the least effective dissuasive factor overall is the threat of unauthorized seizure. While it clearly reduces the nuclear propensities associated with the motive conditions, in many instances these effects are substantially less than those of other dissuasive conditions. Threat of unauthorized seizure, in most instances, produces a drop in nuclear propensity to the moderate region. The motive conditions nuclear threat, overwhelming conventional threat, regional power status/pretensions, and defense expenditure burden are least affected. As can be seen, there are two motive conditions for which no empirical data exist, and so the Bayesian prior estimates are substituted.

Nuclear ally seems to have the greatest overall dissuasive effect. Most nuclear propensities plunge to the bottom of the low region. As one might expect, the effect of nuclear ally is most strongly felt by those motive conditions involving external threat considerations—nuclear threat, latent nuclear threat, and so on—and is least strongly felt by those motive conditions that involve domestic political considerations—for example, domestic turmoil and loss of a war. (This result adds to confidence in both the data and the nuclear propensity method.)

Legal treaty, fear of a preemptive strike by a major nuclear power, and peaceful reputation seem to have moderate dissuasive effects, reducing some nuclear propensities to the weak region and some to the moderate region. Curiously, legal treaty seems least effective in reducing the nuclear propensity associated with regional proliferation. This result suggests that, in the absence of some other dissuasive condition, legal treaty might not be sufficient to contain the eruption of so-called proliferation chains when several motive conditions are present.

Using the values given in table 13, we can calculate aggregate nuclear propensities as outlined previously. Testing the motivational hypothesis in terms of its predictive power is now a relatively simple matter. Based on the definition of a strong nuclear propensity, a proliferation decision should be predicted for every nation-year case whose motivational profile produces an aggregate nuclear propensity of 0.67 or above. When the associated aggregate nuclear propensity value is below 0.67, we predict no proliferation decision. Table 14 presents the resulting tabulation of predictions, along with several statistical tests that gauge the predictive power of the motivational hypothesis.

The chi-square probability gives the likelihood that the apparent association is random. A chi-square probability of 0 means that there is a systematic relation between the motivational variables (as predicted by the motivational hypothesis) and proliferation decisions. Phi is a measure of association that varies from 0 to 1 (reflecting no correlation to perfect correlation). The phi of 0.83 indicates a very strong relation between proliferation decisions and nuclear propensity.

Perhaps the most revealing measures are lambda and tau. These measures gauge the predictive power of the motivational hypothesis by comparing the predictions using nuclear propensity against predictions using "blind guessing." Lambda can be understood as follows: Suppose we were trying to guess whether a specific case represented a decision to begin a nuclear weapons program—say Japan in 1972. In guessing, however, we have no information about the distribution of the cases except that most cases do not involve proliferation decisions. (In fact, 88 percent of the nation-year cases are in this category.) With the intent of minimizing incorrect guesses, we would always guess "no decision," since we know this is what most of the cases are, and consequently we would guess incorrectly only 12 percent of the time. Now suppose we are given additional information: the nuclear propensities for each case. With this additional information, our new guessing rule is: guess proliferation decision whenever nuclear propensity is at 0.67 or above, and guess no proliferation decision when nuclear propensity is less than 0.67. Lambda asks: Does the additional information provided by knowing a case's nuclear propensity reduce the number of errors made in guessing proliferation decisions overall? As an index, lambda measures the corresponding proportional reduction in error against "blind" guessing. A lambda of 0 implies that the additional information does not help in guessing proliferation decisions—that is, there is no systematic relation between the motivational variables and proliferation decisions. A lambda of 0.50 suggests that the additional information cuts the number of guessing errors in half. And, of course, a lambda of 1.0 means there would be no errors in guessing—a perfect relation exists.

Tau, like lambda, is a measure of proportional reduction in errors. Tau, however, assumes that even *more* information is available in making the initial guesses—we are not completely blind in our initial guessing. In particular, it assumes that the actual distribution of the cases with respect to proliferation decisions is known. Thus, in this instance we know that 12 percent of the cases reflect decisions to start nuclear weapons programs, while the rest do not. According to tau, the strategy to be followed would be to predict proliferation decisions 12 percent of the time and predict no proliferation decisions 88 percent of the time. (In contrast, lambda assumed we had no knowledge of the distribution other than that "no proliferation decision" was the modal category.) With the introduction of new information on nuclear propensities, tau asks: Does the additional information provided by nuclear propensity reduce the number of errors made in predicting proliferation decisions over the strategy of guessing the marginal distribution? In essence, both lambda and tau gauge the strength of relation between two variables in terms of the reduction in guessing errors achieved in relation to their respective "random" guessing strategies. In table 14 we see that lambda is 0.78, meaning that knowing nuclear propensity reduces prediction errors (using the lambda blind guessing strategy) by 78 percent. The tau of 0.69 says that predictions based on nuclear propensity reduce the number of guessing errors by 69 percent (over the tau random guessing strategy). Tau can also be interpreted as the amount of variation in the decision variable accounted for (i.e., "explained") by changes in the motivational profiles. That is to say, political and military variables appear to explain about 69 percent of the variation

TABLE 14

PREDICTIVE TEST OF THE MOTIVATIONAL HYPOTHESIS

(NATION-YEAR CASES)

	Prediction	No Prediction
Decision	55 (83)	6 (1)
No decision	11 (17)	450 (99)
Tests		
Chi-square = 138, p = 0		
Phi = .83		
Lambda (row/column) = .78		
Tau (row/column) = .69		

Note: All tests are based on standardized columns. Standardized values appear in parentheses.

in the proliferation decision variable. Both of these give strong support to the motivational hypothesis.

An alternative formulation of this analysis is shown in table 15. Here the cases are nations rather than nation-years. Predictions of proliferation decisions are once again made for cases where the nuclear propensity equals or exceeds 0.67. If a proliferation decision was forthcoming within a year of the prediction, it is considered correct. The reason for this plus or minus one year slack is that errors in both the motivational profiles and the proliferation data are likely to be of the order of one year. The results of the statistical tests are somewhat better than those in table 14. While all the test measures increased slightly (about 10 percent), the changes are not that great. What is important is that the predictive power of the motivational hypothesis is strongly supported by the data, and that two very different formulations of the tests for predictive power

TABLE 15

PREDICTIVE TEST OF THE MOTIVATIONAL HYPOTHESIS

(NATIONAL CASES)

	Prediction	No Prediction
Decision	13 (87)	0 (0)
No Decision	2 (13)	24 (100)
Tests		
Chi-square = 154, p = 0		
Phi = .88		
Lambda (row/column) = .85		
Tau (row/column) = .77		

Note: All tests are based on standardized columns. Standardized values appear in parentheses.

produce roughly equivalent results.[5] Whether we consider "nation-years" or "nations" as the unit of analysis, the statistical results are the same. Strong nuclear propensities are systematically associated with proliferation decisions. The motivational hypothesis is very strongly supported by the data.

ERROR ANALYSIS

While the motivational hypothesis demonstrates very strong discriminating power—reflecting a highly systematic relation between the motivational variables and proliferation decisions—it nonetheless "results in" a number of errors. Since it does view nuclear proliferation as a probabilistic process, one can expect a small number of random predictive errors. Though these errors are relatively few (about 3 percent to 5 percent as displayed in tables 14 and 15), the question remains whether they are random or systematic. If the errors are random, this denotes an inherent level of imprecision in the nuclear propensity model that may remain uncorrected. If, however, the errors are systematic in nature—reflecting either left-out or superfluous variables or a specific deviant country case—then further improvements in the nuclear propensity model are possible.

First, looking across the first row in table 14, we notice that fifty-five (90 percent) of the sixty-one cases representing proliferation decisions are correctly predicted when the unit of analysis is nation-years. Thus the model does very well at detecting prospective proliferants. Of the six cases that are not correctly anticipated by the model, four correspond to the United Kingdom (1949-52). That four of these six errors come from a single country and also are in time order strongly implies systematic error.[6] An examination of Britain's motivational profile during this period reveals that the key variable change was the formation of NATO—that is, Britain entered a formal alliance with a nuclear weapons power. In accordance with the motivational hypothesis—the principle of decision reversibility—Britain should have canceled its independent nuclear weapons program about 1950. The fact is that Britain did not. One explanation invokes the notion of technological momentum. Perhaps once a proliferation decision is made, significant resources are committed, and organizations and institutions are mobilized the elimination of incentives or the appearance of new disincentives may not be sufficient to reverse the decision. There may be a point beyond which the established momentum of a nuclear weapons program simply cannot be overcome by changes in motivational considerations. This in turn implies that there may be a limited "window of reversibility" following an initial proliferation decision during which it is possible to reverse the decision but after which technological momentum becomes the dominant influence driving the program to completion. I will return to this argument when I examine the case of South Korea in the next chapter.

Another possibility concerns the change in Britain's motivational profile in 1950. Britain's nuclear propensity drops from the strong region to the moderate region by 1950 as a result of its entry into the NATO alliance with the United States. The impact of the dissuasive condition nuclear ally is very strong. However, given the American tradition of isolationist retreat after foreign adventures and the British tradition of being the guarantor of security rather than the recipient in its treaty relationships, it is possible that the newly formed NATO did not represent a credible nuclear guarantee (Gowing 1974, 45; Rosecrance 1964b, 59; Pierre 1972, 77). Indeed, United States officials continued to hedge on issues related to the employment of nuclear weapons in defense of Europe. (And, contrary to general perceptions, the United States nuclear weapons stockpile was quite small—no more than about two hundred fission weapons.) Consequently, the dissuasive effect of having a nuclear ally during this early period of the nuclear age may not have been as potent as is generally supposed today. Thus, for Britain the mix of perceived incentives and disincentives may still have strongly favored a proliferation decision.

Turning to the first column in table 14, we can see that eleven of the sixty-six predicted cases of proliferation decisions are incorrect. This 17 percent false-alarm rate, if it reflects random and hence uncorrectable error, is indeed disconcerting. It would mean that the nuclear propensity model can be expected to indicate proliferation decisions erroneously almost one-fifth of the time. More significantly, the error would not be correctable. As it turns out, seven of these seventeen misclassified cases represent a single country, Egypt (1969-80). Once again this is a clear instance of systematic error. The nuclear propensity model indicates that a proliferation decision should have been forthcoming about 1970. Though thereafter its nuclear propensity dropped to the low region, by 1976 it was once again strong, and Egypt should have been actively pursuing an operational capability. However, there is no information publicly available that would allow us to infer that the Egyptian government ever made a proliferation decision.

Among the possible explanations for this systematic error, two are most plausible. The first is that, our technical indicators aside,

Egypt simply did not possess a latent capacity. Owing to some particular quirk in the Egyptian technical, scientific, and industrial resource base, the country may appear to possess a latent capacity according to the technical indicators but actually not be able to support a nuclear weapons program. It must be pointed out, however, that Egypt has been cited by many observers as a "near-nuclear country"—a country assumed capable of such a task.[7] So I am not alone in presuming that an Egyptian government could have made a proliferation decision had it wanted to. The second plausible explanation hinges on leadership variables. Former President Sadat—for reasons of his own—may have chosen not to pursue nuclear weaponry. This explanation is not inconsistent with the motivational hypothesis as long as such instances are the exception. Since the nuclear propensity model, based on theoretical discussions in the literature, tries to formulate some general "norms" of proliferation behavior, it addresses proliferation decision making as it is most frequently observed. In this respect such mispredicted proliferation decisions can be informative and remind us that decision making remains a human endeavor.

To demonstrate how the systematic error of the Egyptian case affects the overall predictive power of the nuclear propensity model, we remove that one country from the data base (i.e., one of forty nations, or seven out of 522 cases) and recompute the predictive measures of table 14. Phi jumps to 0.92, Lambda goes to .91, and Tau increases to 0.85. In other words, viewing Egypt as a deviant case and using the nuclear propensity measure to indicate proliferation decisions reduces our guessing error by more than 90 percent. Where the British case (1947-52) suggests we should reexamine the arguments of technological momentum *once a nuclear weapons program is under way*, the Egyptian case calls for a case-study examination to determine why it deviates from the motivational hypothesis. (Thus quantitative analysis and case study analysis are *not* competing research strategies but are complementary.)

The other major deviant case in this category is Taiwan (1975-76). Taiwan's nuclear propensity is estimated to be strong during 1975 and 1976, indicating a proliferation decision. The conventional wisdom is that, while the Taiwanese government was indeed involved in nuclear option building during this period, no proliferation decision was made. Thus, at face value these indications seem to be in error. However, as I explain in the next chapter, there is equal reason to believe that the conventional wisdom is in error.

Having obtained such gratifying results in testing the motivational hypothesis, there is little reason to continue and examine the null (sui generis) hypothesis. Indeed, it is not possible for the data to be consistent with *both* the motivational hypothesis and the null hypothesis, since the latter postulates a random relation between proliferation decisions and any given constellation of politico-military variables. Had the sui generis model been correct, the Lambda and Tau statistics reported in tables 14 and 15 should have been effectively zero.

APPLICATIONS OF THE NUCLEAR PROPENSITY MODEL

The "quantification" of the motivational hypothesis by the nuclear propensity method does not imply or impart precision above that attainable with systematic qualitative approaches. Indeed, the nuclear propensity method is useful precisely because there is considerable uncertainty and imprecision in the study of political phenomena like the nuclear proliferation process. Where qualitative analyses (e.g., case studies and focused comparison case studies) are particularly useful for illuminating detail and examining specific historical linkages between choice, decision, and behavior, forms of analysis like the nuclear propensity method are well suited for investigating broad patterns of dynamic interactions. The ever-changing nature of the political/military/technical milieu that characterizes the contemporary nuclear era makes the nuclear proliferation process a strong candidate for this kind of study.

In this respect the nuclear propensity method allows us to step back and examine "the forest," since others have already studied the trees in great detail. First we might reexamine the ways technology and politics interact in the nuclear proliferation process. Whether we find general patterns is directly relevant to the problem of devising national and international nonproliferation policies. Specifically, does it make sense to follow a general strategy (e.g., the Nuclear Non-Proliferation Act of 1978, the Non-Proliferation Treaty, etc.), or do we need to devise a specific approach for individual countries? More generally, does it matter if the latent capacity predates strong nuclear propensities—or vice versa—in determining whether a country decides to go nuclear?

Though the aggregate analyses of the relation among nuclear propensity, technology, and proliferation decisions may prove very useful in understanding the nuclear proliferation process, aggregate analyses are not especially illuminating in terms of individual country behavior. Does tracking the nuclear propensities of individual countries over time reveal any interesting patterns? In particular, moderate nuclear propensities are supposed to connote nations "teetering on the brink" of proliferation decisions. Sweden in the early 1960s and Taiwan in the mid-1970s might be interesting to look at. Then there are several known instances of national decisions to acquire nuclear weapons that were reversed before any weapons had been produced. Here India (1965-66) and South Korea (1972-75) invite study. There are also several nations that, sometime in the past, were "accused" of harboring designs on nuclear weapons. The most notable example here is Brazil in the aftermath of the much publicized German-Brazilian Nuclear Technology Agreement of the mid-1970s, and more recently, Argentina. What do Brazil's and Argentina's nuclear propensity values reveal?

At this point, I am in a position to remove the self-imposed restriction to examine only those nations with existing latent capacities. The motivational hypothesis did not require prior latent capacity—only the research design did. The motivational hypothesis merely argued that proliferation decisions were systematically related to specific constellations of politico-military variables. Might nuclear propensities be useful in detecting countries that are indeed interested in "going nuclear" but that first need to acquire the necessary technology and infrastructure? This is where nuclear proliferation routes involving technology purchase, theft, and diversion come into play. Does an examination of Iraq's nuclear propensity reveal anything in the way of its "true intentions" regarding the Osiraq reactor? Does Libya's nuclear propensity reflect its leader's several attempts to purchase nuclear weapons? And does the nuclear propensity indicator reflect Prime Minister Bhutto's 1972 decision to make Pakistan a nuclear weapons power?

It is also time to face squarely the issue of forecasting proliferation decisions in the future. Given subjective assessments of a specific nation's most likely motivational profile over the next several years—and events that might cause significant changes in that motivational profile—nuclear propensities can be estimated. Accordingly, it should be possible to say something about the course of nuclear proliferation over the next several years. And what of the ceteris paribus problem? Might the future be unlike the past? The Bayesian statistical approach introduced in this study has a rather elegant way of assimilating dynamic and structural change into the model. In contrast to most analytic strategies that ignore ceteris paribus, here we can cope with it without trepidation. How we deal with this problem will be developed in the remaining chapters.

6

The Dynamics of the Nuclear Proliferation Process

Both technology and politics are essential ingredients in nuclear proliferation. But exactly how do they interact in setting the stage for proliferation decisions? Here it is helpful to characterize the dynamics of the nuclear proliferation process as one of technologi-cal-motivational convergence.[1] Both the nature of technological capabilities and the relative presence or absence of the motivational variables vary over time. In most instances the progression of technological capability is either stagnant or forward moving. There are, of course, exceptions such as Argentina (as a result of civil strife) and Iraq (as a result of foreign intervention) where the evolution of technological capability is temporarily reversed. However, for most nations, once the latent capacity threshold has been crossed, further technical evolution implies enhancement of nuclear infrastructures and hence increased technical ease in manufacturing nuclear weapons. On the whole, then, the technological aspect of the nuclear proliferation process is fairly monotonic over time. It increases both quantitatively (number of nations with latent capaci-ties) and qualitatively (increasing capability and capacity of individ-ual national nuclear infrastructures) as time goes on.

In contrast, there is no reason to presume a priori that the evolution of motivational profiles will follow a particular trend or pattern over time: there is no reason to assume that nuclear propensities will tend to increase (or, decrease) systematically in the future. Whether the nuclear propensities of individual nations, or nuclear propensities in the aggregate, rise or fall will depend on how the constellations of national and international events and politics develop. As a consequence, any given nation's nuclear propensity may "bounce" between various levels of weak, moderate, and strong. Of course specific motivational variables may exhibit a time trend. Should future cases of nuclear proliferation occur, for example, we would expect the number of nations sensing a nuclear threat to increase. Bear in mind, however, that any future trend in the propensity of governments to consider their nuclear option will be the result of trends in many motivational conditions, not some simple global tendency toward nuclear proliferation.

In the discussion that follows, the nuclear propensity measure is used as a heuristic tool. The purpose is to illuminate some of the dynamic aspects of the nuclear proliferation process that are often missed in more traditional analyses. Nuclear propensity and latent capacity—as more aggregate indicators for motivation and capability—deemphasize the fine detail of history in favor of drawing out general patterns of behavior and interaction. This facilitates cross-case comparisons. The intention is not to replace more traditional analytic approaches (i.e., case studies) but to add to them. Particularly useful in this respect are the nuclear propensity plots for individual countries that show the estimated nuclear propensity in a given year, and a corresponding three-year moving average line as a general trend indicator. These plots are meant to present a broad impression of the ebb and flow of motivational pressures for "going nuclear" in individual country cases. The three-year moving average blends the nuclear propensities of three consecutive years to reflect the more dynamic aspects of the rate and direction of the underlying trend in the balance of proliferation incentives and disincentives confronting a country. In a sense it smooths out the effects of transient events that may affect the indicators of the motivational variables but that may be too fleeting to be perceived as significant by decision makers. By comparing the individual estimated points with the moving average curve, one gets a good impression of how annual changes in a country's motivational profile fall around the more general movement of that nation's nuclear propensity.

TECHNICAL CAPABILITY CONVERGING WITH MOTIVATION

In thinking about the dynamic interaction of technology and politics, consider first the period before a country acquires a latent capacity. Regardless of the nuclear propensity level, nuclear proliferation is not possible because the technological capability does not exist. Yet a strong nuclear propensity—or perhaps even a moderate nuclear propensity—is sufficient to motivate a government to specifically dedicate resources to acquiring a latent capacity; that is, to make a capability decision. Once a latent capacity is acquired, and should strong motivations persist, a proliferation decision—an explicit decision to proceed with nuclear weapons development and production—should follow.

Britain

The case of Britain is particularly illustrative of the convergence of technology with existing motivation because in this instance convergence occurs twice. As is shown in figure 8, Britain's nuclear

Fig. 8. Nuclear propensity of Britain.

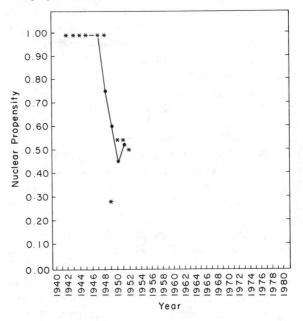

propensity was strong from the very beginning. The principal motive condition was the latent nuclear threat posed by Nazi Germany. German scientists were responsible for many important discoveries related to nuclear fission during the late 1930s. British intelligence was well aware of German research on nuclear weapons. Thus, not only was there a latent nuclear threat, but there was also evidence that the Germans were exploring the possibility of developing what we have termed an operational capability. Britain first acquired a latent capacity by proposing to team up with the United States— taking advantage of United States industrial and raw material capabilities. When technical capability converged with motivation, the proliferation decision was implemented.

When World War II ended in 1945, the United States essentially cut the British off from United States-based nuclear facilities and ongoing nuclear weapons research. In essence, Britain lost its latent capacity. Britain's nuclear propensity remained high, however, as a result of the latent nuclear threat and the overwhelming conventional threat posed by the Soviet Union. Britain's regional and global power status also had important motive effects. In 1947 Britain acquired an independent latent capacity—a direct consequence of the stimulus provided by its enduring strong nuclear propensity. This second convergence of technical capability with motivation gave rise to Britain's second proliferation decision and effort to acquire nuclear weapons.

South Africa

The South African Atomic Energy Board was formed in the late 1940s. Because South Africa was such an important source of uranium ore for the United States and Britain, South African experts were involved early on in discussions related to nuclear science and engineering (Adelman and Knight 1979, 634). Nonetheless, it was not until the end of the 1950s that active research in nuclear science began in South Africa. The Nuclear Physics Research Unit at the University of the Witwatersrand, Johannesburg, was established and plans were made for a research reactor (Spence 1974, 214-15). A subsequent twenty-year agreement with the United States allowed about ninety South African nuclear scientists to visit American Atomic Energy Commission facilities for instruction, training, and research (Adelman and Knight 1979, 634-35). South Africa's Safari I research reactor became operational in 1965. A twenty-megawatt system, fueled by highly enriched uranium, Safari I had no significant plutonium production potential. Its fuel core contained slightly under four kilograms of highly enriched uranium—far too little to be used as the core of a nuclear weapon (Wohlstetter 1979, 169). The entire facility was placed under IAEA safeguards through 1977, which were then extended by agreement to the year 2007. South Africa's second research reactor went critical in early 1968. It was a zero-power critical assembly for experimental nuclear research. Thus, through the late 1960s South Africa had a very modest nuclear research program (i.e., compared with countries such as Brazil, India, Argentina, and Spain) and no latent capacity to produce nuclear weapons.

An important change in South Africa's technological outlook took place, however, in 1970, when the government announced that it was going to finance a pilot-scale uranium enrichment plant. This became operational in 1975, with the capability to produce enough highly enriched uranium to build a dozen or so nuclear weapons a year. Consequently, in 1975 South Africa acquired a latent capacity based on an advanced nuclear infrastructure. With its advanced nuclear infrastructure and its advanced industrial base, South Africa could have had operational nuclear weaponry within two to three years.

What can be said of South Africa's possible interest in nuclear weapons?[2] Figure 9 is a plot of South Africa's estimated nuclear

Fig. 9. Nuclear propensity of South Africa.

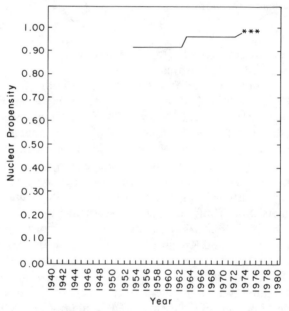

propensity from 1955 to 1980. Up to the early 1960s, the primary motive condition at work was regional power status. The effects of this variable may have been moderated somewhat by South Africa's membership in the Commonwealth—an arrangement that ended with South Africa's withdrawal in 1961. By the early 1960s, the formation of the Organization of African Unity and the United Nations trade embargo increasingly isolated South Africa both on the African continent and in the world at large. South Africa's links to the Western world grew more tenuous. In particular, any notions South African leaders may have harbored about being part of a greater "white commonwealth"—providing diplomatic and security guarantees—were called into question. South Africa's nuclear propensity, already strong, was bolstered when, several years later, a second motive condition entered the motivational profile: pariah status. It was during this later period, 1965-69, that South African officials first began to discuss nuclear weaponry. In 1965 Prime Minister Verwoerd implied that the South African government had a "duty" to consider the military uses of nuclear technology (Betts 1980, 290-92). In 1968 the chief of staff of the South African army noted that South Africa was prepared to manufacture nuclear weapons, while simultaneously calling attention to the country's growing missile program (Betts 1980, 290-92).

Given the strong nuclear propensity that existed throughout the mid-1960s and into the early 1970s, it is conceivable that the South African government initiated a number of efforts specifically aimed at acquiring a latent capacity. South Africa's strong nuclear propensity may have provided the stimulus for a drive to develop a latent capacity that was to be transformed into an operational capability. In this respect the South African case is particularly noteworthy because it reflects a direct jump from a state of no latent capacity to one of latent capacity tied to an advanced nuclear infrastructure. This is, in fact, the technological imperative hypothesis stood on its head. Rather than "technological opportunity" driving political decision making, the South African case illustrates how politico-military motivation may drive technological development. This suggests a "political-motivational" imperative rather than a technological imperative.

In any case, by 1974-75 South Africa's motivational profile received another jolt. A potential "conventional threat" began to form as the Portuguese colonies of Angola and Mozambique gained their independence and joined the African confrontation states. Cuban troops and Soviet military advisers were injected into the region. Though South Africa's nuclear propensity had already peaked, this additional stimulus came at a time when South Africa was about to acquire its (advanced infrastructure) latent capacity. Thus technological capability converged with an already strong nuclear propensity. It was about this time that South African authorities moved ahead with a proliferation decision, later revealed by the nuclear test preparations of August 1977.

Israel

A third and equally demonstrative illustration of technology converging with preexisting motivation is Israel. Israel's first nuclear reactor was a five-megawatt research reactor. Provided by the United States and fueled with highly enriched uranium, it went critical in 1960. It had little utility for producing nuclear weapons usable materials and hence represented no significant increase in Israel's potential to manufacture nuclear weapons (besides providing research data). But the same cannot be said of Israel's second reactor. A natural uranium reactor of French design, it had a nominal power rating of twenty-six megawatts, and when it came on line in 1964 it clearly had the potential to produce usable quantities of plutonium each year—given the availability of sufficient fuel loadings. By virtue of this second reactor, Israel acquired a latent capacity about 1968. Between 1955 and 1976, more than 250 Israeli scientists were trained at United States Atomic Energy Commission nuclear laboratories (recall that only about eighty South African scientists visited United States facilities). Thus Israel came to possess a substantial cadre of trained nuclear specialists. The IAEA claims that by 1977 Israel possessed an unsafeguarded (and unacknowledged) reprocessing plant.

Turning to Israel's nuclear propensity, its time series (not shown) is very similar to that of South Africa. It is consistently in the "strong" region throughout the 1960s. The notable motive conditions in this instance were an overwhelming conventional threat (posed by a coalition of Arab states) and pariah status. Israel first acquired its latent capacity in the late 1960s, a convergence with strong nuclear propensity that resulted in a proliferation decision during this period. Again this is a case where the motivation to "go nuclear" existed long *before* the indigenous technical capability to do so became available. Moreover, historical evidence strongly suggests that Israel's acquisition of a latent capacity was the direct result of a capability decision followed by concerted effort. Israel's continuously strong nuclear propensity provided the impetus for Israeli leaders to fund the purposive development of a latent capacity (Perlmutter 1982). Strong motivation persisted, a proliferation decision followed, and an operational capability was pursued. Again, the dynamics we observe are the technological imperative in reverse.

Motivation Converging with Capability

In contrast to the examples above, there are a large number of cases in which latent capacities have existed for long periods, representing the "constant" in the nuclear proliferation process, but in which motivational profiles have fluctuated considerably. In many in-

stances—especially those where nuclear propensities were weak—
the initial acquisition of latent capacities was the product of basic
industrialization and economic development, not of planned efforts
to create a "nuclear option." In other instances—those in which
nuclear propensities were moderate—capability decisions and nuclear option building may have indeed played a role. In any case, the
more interesting aspect has to do with the convergence of motivation with preexisting technical capability. When nuclear propensities are weak, the acquisition of nuclear weapons should not be a
serious policy issue, regardless of the technical ease with which they
could be produced (i.e., independent of the level of nuclear infrastructure). One might think of Canada, Belgium, and the Netherlands in this category. Yet even weak nuclear propensities might
give rise to nuclear option building *if* there is considerable uncertainty within the government regarding the politico-military environment in the near future. Though the current motivational profile
implies little interest in nuclear weapons, the relative stability of the
variables may be perceived by decision makers as so volatile that
nuclear option building takes place.

When nuclear propensity rises into the moderate region, nuclear
option building is very likely to come into play. Keeping the nuclear
option open—and perhaps even enhancing it by further developing
nuclear infrastructure capabilities—is an important political concern
in the face of a changing politico-military environment and future
uncertainty. Taiwan, France (1948-55), India (1958-64, 1967-71), and
Sweden (1960s) are illustrative.

Finally, as nuclear propensity moves into the strong region, we
expect proliferation decisions. This is the convergence of motivation
with technical capability. The dynamic component here is, of
course, the motivational profile. Since the motivational profile has
significantly more possibilities for variation in time and space, the
dynamics of the nuclear proliferation process in cases where motivations are the primary dynamic component (relative to technical
capability) should be much more volatile.

France

Consider, for example, the case of France (1946-60) as shown in
figure 10. The period 1946-48 reveals a strong nuclear propensity, in
terms both of instantaneous values and of the moving average.
During this period France faced an adversary with a latent capacity
(the Soviet Union), perceived itself as facing an overwhelming
conventional threat (again, the Soviet Union), and was attempting to

Fig. 10. Nuclear propensity of France.

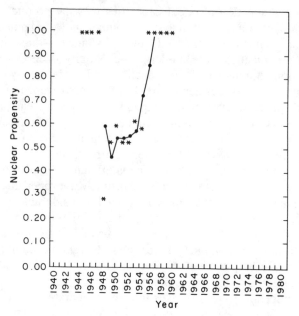

regain the regional and global power status it had held before World War II. The Soviet Union was known to be working to acquire nuclear weapons. While some specialists at the time believed the Russians would require ten to twenty years to duplicate the United States-British feat, others argued that a Soviet bomb was a real possibility by the early 1950s. Whether the Soviet Union would pose a near-term threat to France may not have been too important to French decision makers, since the lead time required to produce a French nuclear weapon was recognized as long.

Though not a military adversary, Britain too was known to be working on its own nuclear weapons. Since Britain was France's most likely regional rival for political and economic influence in Europe, French decision makers also had to consider how Britain's emerging nuclear capability might enhance its influence on the Continent. Could a nation claim regional power status if it lacked the most modern military armaments while its political and military rivals built up nuclear stockpiles?

The French had to confront an analogous issue at the global level. The United States already possessed nuclear weapons, and the Soviets and the British were already several years into their nuclear weapons programs. All of France's foreign policy activities—especially in Asia and Africa—reflected its government's efforts to recover its global position. Could France pursue an independent global policy without nuclear forces, especially given the nuclear weapons programs being pursued by the other global powers? When France finally acquired a latent capacity about 1948—technical capability converging with preexisting motivation—all conditions were ripe for a proliferation decision.

However, a major change in France's motivational profile occurred in 1949. On the one hand, regional nuclear proliferation took place (the Soviet nuclear test), adding a motive condition. On the other hand, NATO was formed, giving France a nuclear ally, the United States. This latter factor accounts for the drop in France's estimated nuclear propensity into the moderate region. (The simultaneous introduction of the motive factor meant that the drop in France's nuclear propensity was not as great as it could have been.) While the conditions were no longer indicative of a proliferation decision, the moderate nuclear propensity that characterized the period 1949-55 was very conducive to nuclear option building.

Between 1949 and 1955, France experienced a number of major political debates over nuclear weapons. There were strong pulls both for and against their acquisition. As is shown in figure 10 France's nuclear propensity hovered within the moderate region, implying an ambiguous alignment of incentives and disincentives. This is precisely the kind of political behavior one should expect with moderate nuclear propensities. Meanwhile, a substantial moderate nuclear infrastructure was being developed. The three Marcoule reactors had some electrical production capability but were even more useful for producing plutonium. Sometime in 1955-56, France's nuclear propensity jumped back to the strong region in response to two changes in its motivational profile. The combination of France's loss of Indochina (war loss) and growing doubts about the credibility of the United States as an alliance partner (especially following the Suez crisis) resulted in a significant increase in France's nuclear propensity. France's proliferation decision was made at this time. The time for nuclear option building was over, and a dedicated nuclear weapons project was approved.

France perhaps best illustrates the distinction between capability decisions related to nuclear-option building and proliferation decisions. Where the former connotes preparation for the option of "going nuclear"—an option that may never be exercised—the latter connotes the decision to move ahead with actual nuclear weapons production. In this respect, had France's nuclear propensity remained strong after 1948, it is likely that technical decisons could have been optimized to produce nuclear weapons several years earlier than the actual first French test in 1960.

India

An equally fascinating case is India, particularly in the combination of its advanced nuclear infrastructure with two independent proliferation decisions—1965 and 1972—separated by a decision reversal in 1966. India's technical development in the area of nuclear science and engineering has been strong since the mid-1950s. Its first research reactor—commissioned in 1956—was built by Indian engineers. Aspara was a pool-type reactor, fueled with highly enriched uranium (supplied by the British). It was CIRUS, India's second research reactor, which went critical in 1960, that offered potential for nuclear weapons production. Of the Canadian NRX design, this forty-megawatt natural uranium system had a significant plutonium production capacity in terms of the base case nuclear weapons program previously discussed. Indian scientists began fuel fabrication for CIRUS, and by 1964 an indigenously designed and built plutonium reprocessing plant came on line. India's nuclear development had proceeded very rapidly: an initial latent capacity by 1958 (no nuclear infrastructure), a moderate nuclear infrastructure by 1960, and an advanced nuclear infrastructure by 1964.[3]

From the motivational perspective, figure 11 shows India's nuclear propensity for 1958-74. From the late 1950s to the early 1960s, India's nuclear propensity was driven by the "latent nuclear threat" and conventional military threat posed by China (with a related conventional threat posed by Pakistan), an unlikely but possible nuclear threat posed by the Soviet Union, and India's own regional power status. The effects of all these motive conditions were moderated by India's self-image as being of "peaceful reputation." India's nuclear propensity was further enhanced by its war loss to China in 1962, and then again in 1964 due to regional proliferation— the Chinese nuclear test.

By late 1965 the combined effect of all the motivational variables began to peak, and, indeed, a proliferation decision soon followed. Prime Minister Shastri gave the go-ahead for a nuclear explosives project that was to culminate in a test detonation sometime in 1968.[4] The subsequent drop in India's nuclear propensity into the moderate region was due to the exit of the latent effects of the 1962 war loss to China and the 1964 regional nuclear proliferation. Mrs. Gandhi, who succeeded Shastri in 1966, correspondingly canceled the nuclear explosives project.

Fig. 11. Nuclear propensity of India.

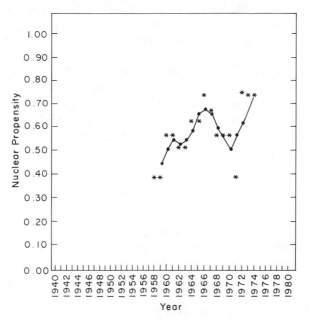

India's nuclear propensity once again jumped into the strong region in the early 1970s. Several changes in its motivational profile account for this. One important addition was the motive condition global power pretensions. As I explained earlier, India's elites began to presume a global role for India, building on its political and technical leadership of the Third World and its newly won politico-military dominance in South Asia (with the dismembering of Pakistan). The eruption of domestic turmoil may also have provided a slight stimulus for the government to look for an undertaking that could symbolize national unity of purpose. And owing to interactive effects among dissuasive conditions, India's signing of the Treaty of Friendship and Assistance with the Soviet Union actually acted to lessen the dissuasive potential of its peaceful reputation. To be sure, many within the Indian political elite did argue that India's tradition-al nonaligned peaceful reputation would be compromised by a treaty with one of the superpowers. It was at this point that motivations again converged with preexisting technical capability. We do know that by October 1972 Mrs. Gandhi approved the design and produc-tion of nuclear explosive devices—one that was subsequently tested in May 1974.

In both Indian proliferation decisions we see motivation converging with preexisting technical capability. These convergence points mark the two proliferation decisions known to have been made by the Indian government. Of equal importance is the effect of motivation *diverging* from technical capability, resulting in a decision reversal. This of course is totally inconsistent with notions of a technological imperative. A proliferation decision had been made, and India's nuclear infrastructure was in advanced stages of development; yet when motivation moved down to the moderate region the decision was reversed. This is further evidence that proliferation decisions may be reversed—if not prevented—by changing national motivational profiles, irrespective of the state of a country's nuclear infrastructure.

South Korea

Another instance of decision reversal—but based on external political intervention rather than on changes in internal decision making—is South Korea. Figure 12 shows South Korea's nuclear propensity for 1965-80. South Korea did not acquire a latent capacity until 1972, so what little activity there was in its motivational profile up to that time is interesting only in contrast to the post-

Fig. 12. Nuclear propensity of South Korea.

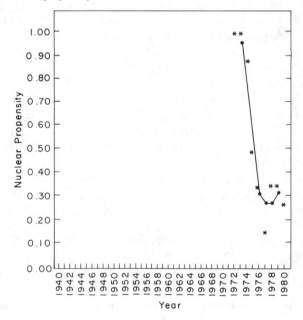

1972 period. The key variable in South Korea's motivational profile is nuclear ally—its defense alliance with the United States. This dissuasive condition minimized the effects of such motive conditions as nuclear threat (China), overwhelming conventional threat (North Korea plus China), and regional nuclear proliferation (China-1964).

However, by the beginning of the 1970s a combination of events led the South Korean leadership to question the credibility of its nuclear ally. First, the Nixon doctrine (United States allies will fight their own battles) and the evolving United States withdrawal from Vietnam suggested a more general disengagement from Asia. Second, there was a growing movement in the United States Congress to bring all American troops home. In late 1971 twenty thousand American troops were withdrawn from Korea. Third, the opening of direct dialogue between the United States and China, the subsequent visit of President Nixon to China, and the Shanghai communiqué which seemed to cast off the United States' former ally Taiwan) may well have led the leadership in Seoul to discount American security guarantees. Thus, about 1972 South Korea's nuclear propensity shot up into the strong zone as the dissuasive influence of nuclear ally was removed from its motivational profile.

Independently, South Korea acquired a no infrastructure latent capacity about the same time. (South Korea differs from Israel in that its latent capacity involved no nuclear infrastructure. It was based solely on underlying industrial and technical capabilities.) It seems that motivation and technical capability converged in this instance to produce a proliferation decision in 1972. Though South Korea did possess a latent capacity, it was a no infrastructure latent capacity that would impose a long lead time to the first bomb, with a relatively small annual production. Moreover, the kind of dedicated nuclear weapons program that would be necessary would certainly attract international attention—perhaps with political and military sanctions. To cut the lead time, and perhaps to provide a cover of legitimacy, representatives of the South Korean government visited Israel, Switzerland, France, and Belgium between 1972 and 1974 to inquire about commercial purchases of nuclear technology. Eventually France and Belgium entered into discussions regarding the construction of a plutonium reprocessing facility in South Korea. In the wake of India's nuclear test in 1974, the Belgians pulled out. Contract negotiations continued with the French.

A particularly interesting aspect of the South Korean case is that the government chose to advance its nuclear weapons program by importing commercial nuclear technologies, though it could have pursued a dedicated nuclear weapons program by indigenous means. A number of possible explanations come to mind. First, a South Korean dedicated nuclear weapons program would have had a substantially smaller nuclear weapons production rate than one based on diversion or conversion of commercial nuclear facilities. Second, a dedicated program would have attracted much unwanted attention very early in the project's life. Both allies and adversaries might have been provoked into responding with political, military, and economic threats, if not actions.

South Korea's nuclear propensity remained strong for several years, until 1975. In 1975 the South Korean government finally ratified the NPT and canceled its nuclear weapons project. South Korea's accession to the NPT was not the cause of the drop in its nuclear propensity, but merely was correlated with it. Actually, the effect of "legal treaty" on South Korea's nuclear propensity is a surrogate for the intense diplomatic pressure that the United States applied in early 1975. Threatening a major break in relations, the United States government took advantage of South Korea's political, military, and economic dependence to compel the South Korean government to cancel its nuclear weapons project. (The United States also forced cancelation of the South Korean-French agreement for a reprocessing facility for South Korea.) The threat of an economic cutoff was particularly potent. By playing on the special dependence of South Korea, the United States was able to produce a rapid change in the incentives and disincentives perceived by the South Korean government.[5] One can only speculate that had the United States succeeded in blocking commercial nuclear sales to South Korea but failed to affect its motivational profile, the South Korean government would have proceeded with a dedicated nuclear weapons program.

And so in examining the cases of France, India, and South Korea we have seen the potential volatility of the nuclear proliferation process. The convergence of motivation with preexisting technical capability—resulting from sudden changes in the politico-military milieu—can give rise to rapid changes in nuclear propensity. The classical admonition that "capabilities change slowly, but intentions can change overnight" is particularly accurate in describing the dynamics of the nuclear proliferation process. Here, "overnight" connotes about a year's time, since in all cases proliferation decisions lagged no more than a year or so behind the convergence of motivation and technical capability.

In the case of France, the duration of the initial convergence (i.e., technical capability with motivation—1948) may have been too brief to produce a proliferation decision—especially when the rather chaotic French political setting at the time is taken into account. But thereafter France's nuclear propensity did hover in the moderate region for several years, providing the impetus for a buildup of it's nuclear infrastructure. With the jump to a strong nuclear propensity about 1956, a proliferation decision quickly followed. France's nuclear option, having incubated throughout the early 1950s, was invoked, and a successful test explosion occurred in 1960.

In the early 1960s, Indian prime minister Nehru had resisted the advice of his scientific-technical advisers to develop nuclear weapons technology. But in concert with its moderate nuclear propensity, he did permit India's nuclear infrastructure to continue to grow and was conscious of its inherent nuclear weapons option. By 1964 India possessed an advanced nuclear infrastructure: plutonium producing reactors, nuclear fuel fabrication facilities, a plutonium reprocessing plant, and a major nuclear science and engineering laboratory. In late 1964 India's motivational profile changed significantly, and within a year a decision was made to invoke India's nuclear option. By 1966-67 another change in India's motivational profile—this time resulting in a downward change in its nuclear propensity—coupled with the rise of new political and scientific leadership produced a decision reversal. Almost as quickly as the proliferation decision had been made, it was reversed. Contrary to the technological imperative hypothesis, a decision reversal—based wholly on domestic political decision making—occurred in the presence of an advanced nuclear infrastructure. Perhaps the decision reversal was facilitated by the fact that no significant technical steps had been taken to implement the proliferation decision (in contrast to the British case). In other words, the "window of opportunity" for a decision reversal had not closed. For the next several years India's nuclear propensity remained in the moderate region. In 1972, however, it jumped back to the strong region; a proliferation decision followed.

The South Korean case is similar. A surge in South Korea's nuclear propensity marked a proliferation decision in 1972. Then diplomatic intervention by the United States—and, in particular, threats of sanctions that took advantage of South Korea's political and economic dependence on the United States—altered the balance of incentives and disincentives perceived by the South Korean leadership. This was reflected in South Korea's motivational profile by its sudden ratification of the NPT—after a delay of six years. A decision reversal occurred, and the nuclear weapons project was canceled.

Contemporary Cases

Thus far the cases I have examined represent fairly unambiguous illustrations of proliferation behavior—that is, there is general agreement on intentions, capabilities, and directions. The analysis of the dynamic interactions of technology and motivations has generally confirmed and reinforced the conventional wisdom. At this point I turn to a number of contemporary cases where proliferation behavior is ambiguous and ill defined.

The cases examined below are Brazil, Argentina, Taiwan, Pakistan, Iraq, and Libya. Brazil is a particularly contentious case concerning the development of a massive nuclear infrastructure. Some people point to this ambitious Brazilian effort to develop nuclear energy as a cover for nuclear weapons development; others argue that it is merely the product of incompetent planning. Argentina is interesting because of its push for nuclear technology autonomy—much like India's effort in the 1960s. Taiwan deserves attention because it is widely believed that its government is interested in nuclear weapons technology and because there have been rumors of military research and development cooperation between Taiwan, Israel, and South Africa. Pakistan, Iraq, and Libya have attracted attention as a result of their disclosed or suspected efforts to acquire nuclear weapons. Pakistan is known to have obtained, surreptitiously, advanced nuclear technologies related to uranium enrichment. Iraq is suspected by many of trying to acquire a latent capacity via possible diversion from its French-built nuclear research facilities. The salience of this case was raised by the Israeli strike on the main reactor at that facility, a dramatic illustration of how suspicion regarding an adversary's latent capacity can provoke military action. Finally, Libya is interesting because of its leader's rumored efforts to purchase nuclear weapons on the "international market."

An examination of the cases of Pakistan and Libya is also warranted because they involve surreptitious efforts to acquire a nuclear weapons capability. Can the nuclear propensity measure be used to "locate" countries that might be engaged in covert activities to acquire a nuclear weapons capability or nuclear weapons themselves? In these two instances we can test whether the nuclear propensity indicator can be used to identify countries that would be motivated to acquire nuclear weaponry if it was possible.

Brazil

Brazil acquired a no infrastructure latent capacity about 1963. We must keep in mind that this reflects an assessment of Brazil's resource base as it pertains to a low-technology, low-output nuclear weapons program. It implies that, had the Brazilian government made a proliferation decision at that time, Brazil could have produced a fission bomb sometime near 1970. However, as figure 13 shows, Brazil's nuclear propensity from the 1960s through the early 1970s was low. The motive condition was regional power status, while the two dissuasive conditions were nuclear ally and threat of unauthorized seizure; the former being the United States (by virtue of the Organization of American States) and the latter as a result of guerrilla activities within the country. By 1967 the threat of unauthorized seizure was contained, as the military government clamped down on domestic political activities.

The contention over Brazil's alleged interest in nuclear weapons dates to 1975, the year the Brazilian-German deal on nuclear technology was signed, estimated to cost between $10 billion and $15 billion. Brazil contracted with West Germany to purchase eight nuclear power reactors (about one thousand megawatts each) and pilot-scale technology for both plutonium reprocessing and uranium enrichment (Cortney 1980). A pilot-scale reprocessing plant would probably be comparable to the Indian facility at Trombay, with an irradiated fuel throughput of one hundred tons a year. This would be

Fig. 13. Nuclear propensity of Brazil.

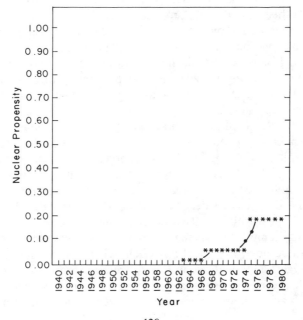

more than ample to produce quantities of plutonium sufficient for a half-dozen or more nuclear weapons a year (depending on fuel burnup). The uranium enrichment facility that was discussed is reported to have had the capacity to produce enough highly enriched uranium for several tens of nuclear weapons per year (Lefever 1979, 109). The sheer magnitude and scope of the agreement—apparently far beyond Brazil's absorption capacity—Brazil's refusal to sign the NPT, and the high output nuclear weapons potential of the facilities led some to suspect that the agreement was a cover for a Brazilian effort to acquire nuclear weapons. Many believed the Brazilian-German agreement was the result of a proliferation decision, not a sincere effort to improve Brazil's energy picture.

Brazil's nuclear propensity did indeed increase about this time to almost 0.20 as global power pretensions entered Brazil's motivational profile. While this is considerably higher than Brazil's previous nuclear propensity, it is still well within the low region. It certainly does not suggest a proliferation decision, or even serious "nuclear option" building. Perhaps one could speculate that the Brazilian leadership had reason to expect a major change in its motivational profile in the future. If we stop to consider potential changes in Brazil's motivational profile—as they might have been perceived by its leader—there is little to suggest a major increase in nuclear propensity in the near future. Brazil has no true military adversary. The political competition with Argentina is not security oriented, and no other power poses any real threat. This rules out the first three motive conditions: nuclear threat, latent nuclear threat, and overwhelming conventional threat. Pariah status seems out of the question. Domestic turmoil might be a possibility and would have the effect of raising Brazil's nuclear propensity to 0.32—perhaps within the moderate region. A loss of war seems highly unlikely given that an interstate conflict involving Brazil is such a remote possibility. Regional proliferation (say, Argentina) would raise the current nuclear propensity only to 0.27; in combination with domestic turmoil it would go to 0.39—a moderate nuclear propensity. Finally, a high defense burden is not likely, since Brazil currently spends only 1 percent of its GNP on military-related projects.

On the dissuasive side, the dismantling of the Rio Treaty—or United States withdrawal from it—would make Brazil's nuclear propensity jump to the strong region. But there was no reason then, nor is there now, to expect any such event. Thus, while figure 13 does suggest a gradual rise in Brazil's nuclear propensity from 1960 to 1980, there is no reason for the Brazilian leadership to have expected any further significant incentives for nuclear weapons to arise (or disincentives to disappear). If building a future "nuclear option" was at all a consideration behind the Brazilian-German agreement of 1975, there is no "motivational" evidence of it. The analysis of Brazil's nuclear propensity suggests no serious Brazilian interest in nuclear weaponry during the period. Perhaps, as the Brazilian government argued, those who pointed an accusing finger at Brazil were overzealous for the cause of nonproliferation.

Argentina

Like Brazil, Argentina has been the target of considerable speculation regarding its "nuclear" intentions. Argentina's nuclear program dates to the early 1950s. The Argentine Atomic Energy Commission came into being in 1950. Traditionally its director has been a military officer, either on active duty or retired, not a civilian scientist—thus attracting international attention regarding the direction of its programs (Cortney 1980). By 1958 Argentina had two research reactors operating: one a low-power critical assembly and the other a 120-kilowatt training reactor of the (United States) Argonaut design. While Argentina's domestic political unrest has had very destructive effects on industry and the economy—and vice versa—its nuclear research program seemingly weathered the turmoil unscathed. By 1967 four research reactors were on line, including a five-megawatt system using highly enriched uranium fuel. A laboratory-scale reprocessing facility also began operating about this time.

What does an examination of Argentina's nuclear propensity reveal about this early period before Argentina had a latent capacity? From the 1950s through the mid-1960s, the primary motive condition in Argentina's motivational profile was regional power status. Argentina had no external security threats, and so the only motive variables relevant to proliferation behavior have to do with domestic politics. Dissuasive conditions present were nuclear ally (by virtue of the Rio Treaty) and fear of unauthorized seizure. Given that, during the period, Argentina's nuclear propensity never rose above 10 percent, there is little to suggest any significant interest in nuclear weaponry or a purposive effort to acquire a latent capacity. One finds little evidence of an effort to quickly acquire the kind of plutonium production capability that India acquired during the same period.

Argentina acquired its latent capacity about 1968. (This apparent lag behind its rival, Brazil, is due to economic and industrial considerations, not problems of science and technology.) By 1969-70 it had a moderate nuclear infrastructure by virtue of its laboratory-scale reprocessing facility—but, in fact, it lacked reactors of significant plutonium production capacity. It was also about this time that its nuclear propensity jumped to about 20 percent, a consequence of domestic turmoil. This motive condition subsided for several years (and nuclear propensity dropped to about 5 percent) only to return in 1974, and it persists today.

More significant are the proliferation implications of the Falkland Islands War. The effects of Argentina's war loss, coupled with its loss of credibility in the military alliance emobodied in the Rio Treaty, have pushed Argentina's nuclear propensity well into the strong region. This convergence of motivation with the capability implicit in Argentina's advanced nuclear infrastructure should see a proliferation decision followed by a first Argentine nuclear weapon operational within several years of such a decision.

Taiwan

In the mid-1970s Taiwan attracted the attention of proliferation watchers as reports surfaced of secret plutonium reprocessing activity at Taiwan's main nuclear research laboratory. In contrast to the South Korean case, which occurred at about the same time, most reports of official (though unattributed) United States reactions suggested that Taiwan was not actively engaged in a nuclear weapons project (Yaeger 1980, 79; Binder 1976, 6). Instead, its activities were generally interpreted as nuclear option building. For this reason the Taiwan episode was not recorded in this study as a proliferation decision. Strong United States pressure eventually caused the reprocessing facility to be shut down and brought a statement by the Taiwanese leadership disavowing interest in nuclear weapons.

What does Taiwan's nuclear propensity time series reveal about this period? In figure 14 we see that throughout the 1960s Taiwan's nuclear propensity remained low, largely owing to the dissuasive influence of its nuclear ally, the United States. Taiwan's foremost nuclear research center was established in 1968, and subsequently Taiwan ratified the NPT, resulting in a further reduction in its nuclear propensity. Taiwan subsequently acquired a latent capacity in 1969. In 1972 its nuclear alliance with the United States was called into question as United States-China relations warmed; culminating in the Shanghai communique. This was followed by the withdrawal of United States military forces from the island. By 1973 Taiwan's TRR research reactor came on line—a forty-megawatt natural uranium reactor similar to India's CIRUS. Taiwan now possessed a

Fig. 14. Nuclear propensity of Taiwan.

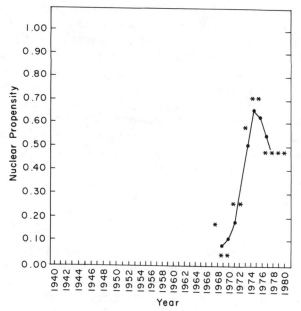

moderate nuclear infrastructure capable of producing enough plutonium for at least one nuclear weapon a year. The TRR cut Taiwan's
lag time to an operational capability from six to four years. This
might not be noteworthy except that within a year Taiwan's nuclear
propensity jumped to the moderate region as the result of regional
proliferation (i.e., India) and then into the strong region in 1975 as
Taiwan attained pariah status (Yaeger 1980, 79-80; Binder 1976, 1).

At this point Taiwan should have made a proliferation decision.
Indeed, it is at this point that Taiwan is reported to have begun the
secret reprocessing of plutonium. As I noted, this activity was
interpreted by many as nuclear option building, implying that no
proliferation decision had been made. Yet the nuclear propensity
level suggests that conditions were in fact ripe for a proliferation
decision. Here the analysis contradicts the conventional wisdom:
though I did not include the Taiwan episode as a proliferation event
in my examinations in chapters 4 and 5, the motivational model
argues that it should have been there. It is indeed possible that in
1974-75 the Taiwanese government decided to push ahead with the
production of nuclear weapons. (This would be the first clear
evidence of a state's violating the NPT, an event the United States
government might prefer to keep secret.)

By 1977-78, however, the effects of regional proliferation subsided, and Taiwan's nuclear propensity dropped into the moderate region. The reprocessing activity was halted by United States diplomatic pressure and the threat of economic sanctions. If we accept the indications of the nuclear propensity measure—that 1975 saw Taiwan's leader make a proliferation decision—then this halt to Taiwan's reprocessing represents a decision reversal. If not, then the United States intervention merely kept Taiwan from enhancing its nuclear option and further reducing its lag time in the future. In either case, Taiwan's nuclear propensity remains in the high-moderate range and any one of a number of motive conditions could kick it back into the strong region. Another instance of regional proliferation, for example, would provide the impetus for Taiwan to restart its efforts.

Pakistan

If anything demonstrates how strong nuclear propensity can drive a country to pursue every possible avenue to develop the capability to produce nuclear weapons, it is the case of Pakistan. Figure 15 gives Pakistan's nuclear propensity time series for 1960-80. As is shown, its nuclear propensity remained in the weak region through 1970.

Fig. 15. Nuclear propensity of Pakistan.

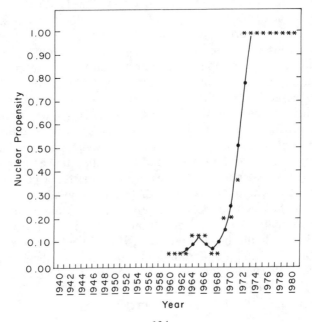

The latent nuclear threat posed by India (and a possible nuclear threat posed by the Soviet Union) were both moderated by Pakistan's alliance relationship with Britain and the United States. The jump observed in 1964 was the result of regional proliferation, while the jump in 1969 was due to domestic turmoil.

Pakistan did not possess a latent capacity during this early period and did little to develop its nuclear infrastructure. Though its first nuclear research center was founded in the mid-1950s, its first research reactor did not come on line until 1965—a five-megawatt reactor that used highly enriched uranium fuel, but of a design with little value for producing material for nuclear weapons. By late 1972, however, a 125-megawatt power reactor came on line that was capable of producing enough plutonium for several nuclear weapons a year. Yet even with this potential source of plutonium Pakistan did not acquire a latent capacity, because it lacked the indigenous capability to build a reprocessing facility. Thus, up to the early 1970s Pakistan appeared to lack both the capability and the motivation to "go nuclear." With the loss of East Pakistan (war loss) in 1971 Pakistan's nuclear propensity jumped to the moderate region. Correspondingly, the unwillingness, or inability, of the United States to prevent the dismemberment of its ally resulted in an understandable loss of alliance credibility, and hence nuclear ally dropped out of Pakistan's motivation profile. The consequence: a significant jump in 1972 to a "strong" nuclear propensity, where it has remained. Lacking a latent capacity, we would expect to observe—as in South Africa and Israel—a capability decision by the Pakistani government to develop a latent capacity. And in fact there is substantial evidence that this did indeed occur. Pakistan's former prime minister, Bhutto, is reported to have approved, in 1972, a wide-ranging series of activities designed to give Pakistan the basic capability to manufacture nuclear weapons—to acquire a latent capacity.

Significantly, the Pakistani government planned a direct jump from no latent capacity to advanced nuclear infrastructure. Its initial strategy was to procure a plutonium reprocessing plant from France. While negotiations between the French and Pakistanis proceeded, the Canadians cut off further fuel supplies to the Pakistani power reactor pending accession to the NPT. At this point Libya's Qaddafi offered to provide financing and uranium to Pakistan in exchange for several nuclear weapons in the future. Bhutto apparently agreed. By 1979, however, the French had begun to have second thoughts about their aid to Pakistan and backed out of the agreement, leaving the reprocessing plant unfinished.

Even before this point Pakistan had begun a clandestine effort to acquire the components for a centrifuge uranium enrichment facility. As Western efforts to close off Pakistan's access to the necessary components for completing the centrifuge facility increased, Pakistan redirected its attention to completing the reprocessing plant. Work also continued on the enrichment facility. Either way, the eventual acquisition of a latent capacity tied to an advanced nuclear infrastructure will put Pakistan within several years of nuclear weapons production.

The Pakistan case also illustrates the importance of the "Sears, Roebuck" and "Yellow Pages" factors mentioned in chapter 2. Pakistan still falls short in some critical areas of a no infrastructure latent capacity—it remains incapable of producing nuclear weapons using all indigenous means (in contrast, for example, to South Korea). Indeed, the weaknesses in Pakistan's resource base are reflected in its inability to complete its reprocessing facility and to keep its uranium centrifuges running. Annual forecasts since 1976 that Pakistan would have the bomb within two years (1978, 1979, 1980, etc.) thus have consistently been wrong. To date Pakistan lacks a latent capacity. What the case of Pakistan amply demonstrates is the extent to which strong motivation to "go nuclear" can push a nation to acquire the necessary technology. Thus, as with South Africa and Israel, this is the technological imperative in reverse.

Iraq

The controversy over Iraq's nuclear intentions was simultaneously made more acute, but less significant, by the Israeli strike against Iraq's nuclear facilities in June 1981. The target—the Osiraq reactor—was designed to use highly enriched uranium fuel, generating a peak power of about 70 megawatts. Owing to its highly enriched uranium fuel, this reactor could not produce meaningful quantities of plutonium within its fuel loading. Instead, its potential utility to an Iraqi nuclear weapons project would involve either direct diversion of its enriched uranium fuel *before* it was placed in the reactor (to make a uranium-based weapon), or modifying the reactor shield design to incorporate a natural uranium "plutonium-breeder blanket." In the latter instance the blanket of natural uranium would capture neutrons during reactor operation and undergo conversion to plutonium. The blankets would then have to be removed from the reactor and reprocessed to remove the plutonium.[6]

In the former instance—direct diversion of the enriched uranium

fuel—proper circumstances might theoretically have found sufficient quantities of highly enriched uranium available for several
nuclear weapons. To be sure, any such diversion would have led to a
cutoff of reactor fuel supplies from the French, but the "utility" of
several nuclear weapons and the lure of a simpler nuclear weapon
design might have made this appear a worthy option.

The Israeli strike, which demolished the reactor about six months
before it was to become operational, can thus be seen as an effort to
prevent a latent capacity threat from developing. The Israeli leadership was convinced that Iraq intended to acquire nuclear weapons.
But is there any evidence of Iraqi motivation to explicitly develop a
latent capacity and transform it into an operational capability?
Figure 16 shows Iraq's estimated nuclear propensity from 1965 to
1980. Generally speaking, we see that Iraq has had a relatively
volatile motivational profile over the past decade and a half. This
has resulted in a nuclear propensity that has bounced from strong to
weak to moderate, and so forth. In this respect it is somewhat
reminiscent of the French nuclear propensity time series during the
early 1950s. A major difference, of course, is that Iraq lacked a
latent capacity.

The peak in Iraq's nuclear propensity (1968) is the result of the

Fig. 16. Nuclear propensity of Iraq.

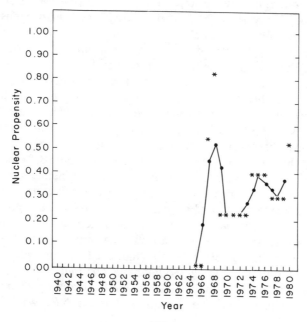

latent nuclear threat posed by Israel and the defense burden that Iraq began to carry at that time. The following year Iraq ratified the NPT, and its indicated nuclear propensity dropped to the low region. During this time Iraq appeared to do little toward developing a latent capacity—behavior consistent with its generally low nuclear propensity. Its newly acquired two-megawatt research reactor was not the scene of much activity. By 1974, however, we see Iraq's nuclear propensity move up to the moderate region—a consequence of regional proliferation. Regional proliferation effects developed on two levels. On the one hand, India detonated its nuclear device, while on the other hand a number of reports surfaced strongly arguing that as of 1973 Israel possessed operational nuclear weapons. Consistent with its moderate nuclear propensity, Iraq initiated a number of actions aimed at acquiring nuclear technology and, correspondingly, a latent capacity (Winkler 1981).

In 1974 Iraq concluded a nuclear cooperation agreement with India and began negotiations with the French for nuclear technology. Although the Indian link never developed, Iraq and France did conclude an agreement for nuclear cooperation—part of which included the seventy-megawatt Osiraq reactor.[7] Also at this time, the Iraqis approached the Italians regarding the purchase of "hot cells"—facilities that could be used for plutonium reprocessing. Agreement was reached in 1978 for a facility of sufficient capacity to produce from one to two nuclear weapons per year (Winkler 1981; Rowen and Brody 1980, 207). Though Iraq's nuclear propensity at this point dropped just below the moderate level (as regional proliferation dropped out of the motivational profile), the decrease was so slight as to be insignificant. Moreover, by 1980 Iraq's nuclear propensity jumped well up into the moderate region, as the dissuasive effects of threat of unauthorized seizure dissipated. Then in 1982 Iraq's nuclear propensity jumped to the strong region as a result of its war loss to Iran and the combined, but independent, conventional threats posed by Israel and Iran.

In sum, there is evidence in the motivational profile data that Iraq was interested in developing a *nuclear option*. Whether its leader would have made a proliferation decision once Iraq acquired an initial latent capacity would depend on the motivational profile at that time. Assuming that Israel had not attacked and destroyed the Osiraq reactor, and that Iraq's motivational profile remained as it was in the beginning of 1982, the indicated nuclear propensity implies that option building would have continued, but that no proliferation decision would have been forthcoming. This also assumes, of course, that Iraq's signature on the NPT was sincere and not just a cover to acquire nuclear technology. That Iraq signed the NPT while its nuclear propensity was strong, and that there were no accompanying political or military events to compel such an action (say, in contrast to the South Korean case), gives reason for reflection. On the one hand, if Iraq's ratification of the NPT was indeed a ruse, then its nuclear propensity, in reality, has been continuously in the strong region. This implies both a capability decision and, if conditions persist, implementation of a proliferation decision.

Libya

The case of Libya is, perhaps, as bizarre as is its leader, Colonel Muammar Qaddafi. Since coming to power in a coup in 1969, Qaddafi has become well known for his mercurial behavior. There have been persistent rumors of his trying to obtain nuclear weapons through purchase, theft, and international collaboration. To be sure, given Libya's complete lack of nuclear infrastructure—not possessing even a research reactor—an "unconventional" strategy for acquiring nuclear weapons would be necessary.

Allegedly, sometime in 1969 or 1970 Qaddafi approached the Chinese about buying one of their nuclear weapons. The Chinese premier, Chou En-lai, reportedly refused (Cooley and Kaufman 1980). Several years later Qadaffi turned to Pakistan's leader Ali Bhutto and offered to support the Pakistani nuclear weapons effort. Flushed with money from the rise in the price of oil, Libya would provide funding and uranium if Pakistan would share the resulting nuclear stockpile (*Washington Star* 1981a). Examining Libya's nuclear propensity up to 1975 (fig. 17), we see that it was strong. The most important motive condition was the latent nuclear threat posed by Israel. Moreover, Qaddafi's numerous attempts to merge with neighboring states suggest some pretensions to becoming a regional power. So his efforts to acquire nuclear weapons are consistent with Libya's nuclear propensity.

Meanwhile Libya signed the NPT (1975), most likely at the insistence of the Soviet Union, with whom Libya was negotiating for a four-hundred-megawatt power reactor. Given Libya's motivational profile (and strong nuclear propensity) just before its ratification of the NPT—not to mention circulating rumors of Israeli nuclear weapons—Qaddafi may have had to sign the NPT to get direct access to nuclear technology. In other words, there is every reason to discount the role of legal treaty in Libya's motivational profile. For all intents and purposes, then, Libya's nuclear propensity

Fig. 17. Nuclear propensity of Libya.

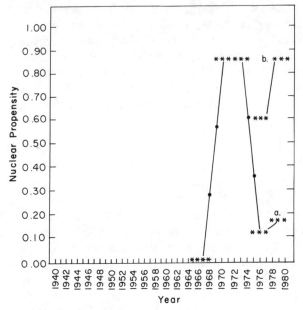

a. Including "Legal Treaty"
b. Excluding "Legal Treaty"

should be estimated without the dissuasive effects of legal treaty. (Fig. 17 shows the effects of both including and excluding this dissuasive factor.) The Soviets apparently have had similar concerns, since they have failed to proceed with the power reactor program.

Up to 1979 Qaddafi had generously financed the Pakistani nuclear program and provided more than 450 tons of uranium—purchased from Nigeria (*Washington Star* 1981a, b; Cooley and Kaufman 1980). With the overthrow of the Bhutto regime, Qaddafi became concerned that the Pakistanis would not fulfill their part of the deal and is reported to have insisted on participation of Libyan technicians. The Pakistani leadership refused, and Libya withdrew its support (*Washington Star* 1981a).

Unable to purchase commercial nuclear facilities or buy into the Pakistani program, Qaddafi turned once again to the international marketplace and offered somewhere between $100,000 and $1 million to anyone who would deliver to him an operational nuclear weapon. There were reported to have been at least two responses, and apparently Qaddafi actually paid for a bogus bomb (*Washington Post* 1980). From figure 17 we see that Libya's nuclear propensity at this point was in the moderate range if the dissuasive effects of legal treaty are included, or it was in the strong range if legal treaty is discounted.

SUMMARY

The results of the dynamic analyses suggest that some general observations are in order. The cases of Britain, South Africa, Israel, and Pakistan illustrate the inherent danger of ignoring countries that have strong nuclear propensities but lack the technical capability to manufacture nuclear weapons. Should the motivational profile remain substantially unchanged, then in time technological capability will surely converge. Even without an international market for nuclear technology, national economic development will provide the rudiments of a latent capacity. Moreover, as we have seen, a strong nuclear propensity actually stimulates government investment toward creating a latent capacity. In the end this convergence of technological capability with preexisting motivation results in nuclear proliferation. The Libyan case further attests to the proliferation threat posed by strong nuclear propensities in the absence of technological capability. Qaddafi's past failures to purchase ready-made nuclear weapons notwithstanding, the odds of success may improve in the future should additional states acquire nuclear weapons—states lacking in secure command and control.

The cases of Brazil and Argentina, as well as others not considered, such as Sweden and Switzerland, demonstrate how emphasizing technical capability over motivation can lead one astray. In the wake of the Indian nuclear test of 1974, political sensitivity regarding the threat of nuclear proliferation increased. Much of the writing produced by researchers and analysts outside government focused on those countries where large nuclear energy programs seemed in the offing—but whose nuclear propensities were low. In other words, the indicator of potential threat of nuclear proliferation was capability, not intention. In the meantime, countries such as South Korea and Taiwan were quietly moving ahead with dedicated efforts to develop operational capabilities. Public attention even turned away from Pakistan once the French-Pakistani deal for a reprocessing facility was canceled. The Pakistanis, however, surreptitiously acquired the components for a nuclear centrifuge plant while initiating indigenous efforts to bring their own reprocessing plant on line. Only when it was (again) apparent that Pakistan might possess the technical wherewithal to "go nuclear" did it once again receive widespread international attention.

All these cases are suggestive of what I have termed the "motivational imperative." Trends in science, technology, and economic/industrial development are such that in the long run most states will acquire a basic latent capacity to produce nuclear weapons. Thus, unless something occurs to alter motivational profiles, all states with strong nuclear propensities will eventually "go nuclear." The convergence of technical capability with motivation may be a long drawn-out process, but we have seen that many of the motivational variables are stable enough to persist over a decade or longer.

The cases of France, India, South Korea, and perhaps Taiwan demonstrated the potential volatility of motivational profiles. Nuclear propensities can, and do, change radically within short periods. Having noted two or three decision reversals, we see that these cases present strong testimony to the pivotal role that politico-military variables play in proliferation decision making. And there are also notable instances of motivational ambiguity, where the balance of incentives and disincentives is such that a proliferation decision is neither compelled nor unlikely. Historically, nuclear option building seems to be the mode of behavior for countries with moderate nuclear propensities. But this should also be interpreted as uncertainty on the part of decision makers, who have reason to believe that in the near future incentives may outweigh the disincentives and prompt a proliferation decision. Nuclear option building, in turn, will shorten the lag time to an operational capability.

Thus we must return to the relative roles of technology and politics. It should be obvious that each is necessary but that only both together are sufficient for proliferation decisions. The actual cause of proliferation decisions is politico-military motivation. It is the relative balance of politico-military incentives and disincentives—as presented by the prevailing constellation of motivational variables—that weighs on the minds of national decision makers. When the incentives dominate the disincentives, decision makers will, more often than not, opt to convert latent capacities into operational capabilities. If a latent capacity does not exist, effort will most likely be directed toward acquiring one.

Technology is not a cause of nuclear proliferation, it is an aid. Technology provides the opportunity to implement proliferation decisions, and it may even affect the balance of incentives and disincentives. As a nation's latent capacity evolves through the various stages from no infrastructure to advanced infrastructure, the financial and resource costs of "going nuclear" obviously will decrease. Moreover, the potential to produce large quantities of nuclear weapons is likely to increase. That is to say, more advanced nuclear infrastructures, especially those based on civil nuclear power programs, are likely to produce greater amounts of bomb-usable material than would a first-effort dedicated program.

This last argument has led some to argue that technology may in fact alter the balance of incentives and disincentives. When the financial and resource demands of "going nuclear" become less burdensome, states might opt to proceed with nuclear weapons production under a balance of incentives and disincentives that traditionally might have been perceived as insufficient for a prolif-eration decision. In other words, technological plenty could lower the nuclear propensity threshold. I will explore this issue in detail in the next chapter.

Before concluding, it is worth returning to the Libyan case, since it accents an important point of methodology. The Libyan case, more clearly than any of the others, demonstrates the importance of historical data in applying the nuclear propensity method. Behind each motivational variable there exists a set of simple, but nonethe-less important, assumptions. For example, the dissuasive condition legal treaty assumes that accession to a treaty is a sincere undertak-ing. It is presumed to be a surrogate for some other condition, or set of conditions, implying little or no interest in nuclear weaponry. Here a comparison of the actual reasons why the Netherlands, Canada, South Korea, Taiwan, Iraq, and Libya ratified the NPT might prove instructive.

The point is that historical data and analysis are necessary to a valid application of the nuclear propensity measure. Again, it was not introduced here as an alternative, but rather as a different method of historical analysis. While history is interesting in itself, one would like to believe that a better understanding of history would enable us to manage contemporary affairs in a better manner. In particular, where public policy is concerned, knowledge of the past may prevent us from repeating mistakes. Though asserted less often, it may also aid us in avoiding new blunders. What remains to be explored in the final chapter is the utility of the motivational model for addressing nuclear proliferation as a contemporary policy problem. In this regard, one obvious avenue of pursuit involves assisting policymakers in anticipating current and developing trends in the nuclear proliferation process: forecasting.

7

Forecasting Nuclear Proliferation

Forecasting has been a part of the literature of nuclear proliferation since the first United States government memorandum on the spread of nuclear weaponry was written in the mid-1940s. How many countries will have the capability to produce nuclear weapons by the end of the decade? How many nuclear weapons powers will there be by 1990? Which countries are they likely to be? What will be the global production of nuclear energy in the year 2000, and what does it imply for the level of worldwide plutonium stocks? Looking back over the works in this field, it quickly becomes apparent that practically all analyses of the problems and risks associated with nuclear proliferation are written in the future conditional. Indeed, within the policy process, the role of forecasting has been to agitate policymakers into action by pointing out that, while the threat of nuclear proliferation (i.e., a multinuclear world and the risks inherent in it) has yet to fully materialize, it is looming just over the horizon. Besides maintaining a modicum of interest in the subject, however, forecasting has done little else to assist in the formulation and implementation of government policy toward nuclear proliferation.

Like forecasts of general stock market trends, forecasts of the general trends of nuclear proliferation—for example, the number of nuclear powers in 1990—are interesting but of little practical use. To be sure, they alert us to the possible risks and dangers ahead and the consequent need for action. But such forecasts fail to provide any information to aid in decision making. In contrast, a forecast of how a particular stock is going to behave under given conditions, what effects changes in that stock will have on other specific stocks, and how long investors will have to react would prove most useful. Can

nuclear proliferation forecasts offer equally useful information to
policymakers?

The concept of forecasting introduced here involves far more than
the mere prediction of events and trends. Think of nuclear proliferation as an infectious disease—one we would prefer to limit and
isolate, if not prevent. Forecasting for public health and disease
control is required to do more than predict the total size of
epidemics. The population at risk must be defined and stratified in
terms of relative susceptibility. In our case the task is determining
the relative likelihood that countries will make proliferation decisions and subsequently acquire nuclear weapons within a fixed time
frame. The results of the previous chapters show that, from a
motivational perspective, some countries clearly "deserve" closer
watching than others. Of equal importance is to determine whether
there are some subgroups within the population at risk that might
suffer more severe consequences than others or pose greater haz-
ards to the general welfare. For example, some countries are more
likely to be placed in a situation where the use of nuclear weapons is
possible—others are more likely to cause further proliferation after
their own acquisition of nuclear weapons. (Some countries have a
greater likelihood of producing contagionlike effects.)

These two factors, the likelihood of going nuclear and the
consequences resulting from it, define the risk associated with any
given country's acquiring nuclear weapons. However, from the
perspective of nonproliferation policy, two additional factors are
relevant. Continuing with the health services analogy, forecasting
must also consider the options for, and potential effects of, intervention and treatment. Of particular importance are variations in
sensitivity to treatment within the population at risk and the time
available for intervention and treatment. Nuclear proliferation forecasting likewise can address the relative responsiveness of countries
to antiproliferation efforts and give some indication of the relative
length of time available to produce decision reversals (i.e., before
nuclear weapons are actually produced).

Thus we have identified four components of risk that forecasting
could aid policymakers in evaluating: likelihood of specific countries' going nuclear, expected consequences, probable responsiveness to antiproliferation efforts, and time available for antiproliferation efforts.

THE LIKELIHOOD OF NUCLEAR PROLIFERATION AS A RISK
FACTOR

Likelihood is an obvious risk factor that does not require much discussion. The incentives and disincentives related to proliferation decision making vary considerably between countries and over time, and, as I have shown, so does corresponding interest in acquiring nuclear weapons. Forecasting likelihood gives the policymaker some prior notion of the *chance* that a given country will come to be a proliferation problem—that is, the prospects that it will pose some level of risk to nonproliferation goals.

Previously, most efforts at forecasting likelihood involved asking experts for their impressions of the probability that specific countries would go nuclear within some fixed time frame. More sophisticated designs incorporated alternative future events and then asked for conditional probabilities. The problem with these aggregate approaches is that there is no way to accurately assess the weights and combinatorial rules that the various experts employed in reaching their net assessments. It is difficult to determine what factors different experts included or excluded, and therefore it becomes difficult to bring their judgments together.

However, the nuclear propensity indicator does offer an alternative for forecasting likelihood. First, experts are asked to make probabilistic statements regarding the likelihood that specific politico-military variables (as described in chap. 3) will be perceived as present or absent by a given country's leaders over a fixed period. Many such assessments for each variable result in a probability distribution. Then, using the empirical relationships that associate specific arrays of politico-military variables with overall nuclear propensity, a resulting probability distribution for each country's nuclear propensity can be generated (for a given time frame). Thus, experts are being asked to judge individual components, which can then be combined systematically into an estimate of nuclear propensity. Finally, so as not to misrepresent the precision of the measure, the three-level ordinal scale for nuclear propensity (weak, moderate, high) can be used.

Too often, likelihood has been used as the only factor for characterizing the risk inherent in a country's acquiring nuclear weapons. Though not obvious at first, implicit in this approach is the presumption that all instances of nuclear proliferation are of equal consequence—that is, all are equally undesirable. If the consequences of each occurrence of nuclear proliferation are independent of the specific country involved, then likelihood is indeed the appropriate measure of risk. Forecasts of aggregate trends in nuclear proliferation—for example, the future size of plutonium stocks, or the probable number of nuclear weapons states—do imply some equivalence between quantity and threat (risk). More trend is more risky, and the composition or distribution of countries that make up the trend is of little importance.

Yet intuitively the assumption of equivalence of consequences does not hold. If forced choices are posed, students of nuclear proliferation show consistent preferences. For example, most would prefer to see Canada acquire nuclear weapons rather than Pakistan, Australia rather than the Federal Republic of Germany, and Brazil rather than Libya. Other forced choices might not yield obvious preferences, but a fairly strong preference ordering can be found across the full range of cases. Thus nuclear proliferation forecasts must treat explicitly the expected consequences associated with an instance of nuclear proliferation.

PROLIFERATION SALIENCE AS A RISK FACTOR

In the context of nonproliferation policy, salience reflects the perceived importance of a specific country's continued adherence to the nonproliferation norm. It is a measure of the expected consequences of failure to prevent a given country from going nuclear. As a risk factor, salience aids the policymaker in determining the relative priority for attention to be assigned to prospective proliferants when only the consequences of proliferation are considered.

Strictly speaking, any attempt to estimate the relative salience levels of a sample of countries would require a fairly explicit delineation of the objectives of the prevailing nonproliferation policy. For example, if the primary objective is merely to prevent any incidents of nuclear proliferation during a given administration's tenure in office, then salience would be synonymous with near-term nuclear propensity. (Any instance of nuclear proliferation would be equally damaging to the policy goal.) If the primary objective is to ensure that no direct military adversary of a United States ally goes nuclear, salience might be judged as some combination of nuclear propensity and geographical proximity to United States allies. Thus salience, to the extent that it can be considered a property of a country, can be judged only in the context of the objectives of nonproliferation policy.

In a more general sense, however, the objectives of nonproliferation policies have been linked to two main concerns: (1) impact on interstate politico-military instability and conflict and (2) impact on further nuclear proliferation. When people speak of the "risks" of nuclear proliferation, it is these two threats that surface. Further nuclear proliferation might upset fragile political and military balances. Possibilities for conflict involving nuclear weapons would increase, as might other countries' interest in going nuclear. (So the expected consequences of one country's going nuclear might be to increase the likelihood that others will make proliferation decisions.) Terrorists and other nonstate actors might find it easier to acquire nuclear weapons through theft or purchase, or as gifts.

To measure salience, I surveyed a dozen students of nuclear proliferation. Those specialists were asked to categorize each country as weakly salient, moderately salient, or strongly salient. The criteria for judging salience were as noted above: expected impact on regional stability and conflict and on further proliferation. For forecasting purposes, the median value was assigned as the best estimate of a country's true salience (though agreement in these judgments, as might be expected, was very high).

TREATABILITY AS A RISK FACTOR

The third factor is treatability, the degree to which a given country would be responsive to antiproliferation intervention. How difficult would it be to reverse a specific government's chosen course to acquire nuclear weapons? In part, treatability is a reflection of a country's individual political, military, and economic relations with the outside world. By virtue of their particular economic dependence on the United States, Taiwan and South Korea both were highly susceptible to United States pressures to back away from nuclear weapons development during the 1970s. While Pakistan may also have been in this category during the 1960s, its newfound benefactors in the Islamic world have greatly reduced its dependence on the United States. Other states are in analogous situations with respect to the Soviet Union—Cuba, for example. In a crude sense, treatability gauges how easy it would be to bribe, extort, cajole, embarrass, and coerce a country into reversing a proliferation decision.

Yet if treatability reflects "the offer a prospective proliferant cannot refuse," it also reflects the lengths to which the managers of the nonproliferation regime are willing to go in the name of their nonproliferation policy. Iraq, for example, was seen as much more treatable under the action guidelines of Israel's nonproliferation policy (which included military action against dangerous facilities) than it was under the French policy. Treatability is also affected by the general state of political relations between countries. Under the Carter administration the politics of human rights strained relations with a number of states; for example, Argentina. Before the Falkland Islands crisis, the Reagan administration had succeeded in breathing new life into relations between Argentina and the United States. One might guess that Argentina's treatability rose accordingly. But it is also possible that the Reagan administration would not pursue antiproliferation efforts with the same vigor as the Carter administration. Accordingly, Argentina's treatability might have decreased. In any case, the renewed friction between the United States and Argentina after the Falkland Islands crisis portends a further decrease in Argentina's treatability.

Given all the uncertainties, variables, and subjective aspects involved, a survey approach seems appropriate for forecasting treatability. Specialists could be polled and asked to rank each country in terms of its relative treatability (low, medium, high) under "foreseeable" United States nonproliferation policy. However, this will be a particularly difficult task because two complex judgments are necessary: (1) how far nonproliferation policy would be pushed to reverse a given proliferation decision and (2) how responsive that country might be.

THE LAG TIME RISK FACTOR

Another variable that has often been used to denote risk in studies of nuclear proliferation is lag time. Lag time is the amount of time a given country requires after the proliferation decision to produce its first nuclear weapon. Countries such as Japan and Canada have very short lag times—of the order of months, if not weeks. Other countries—for example Paraguay and Gabon—have extremely long lag times: of the order of decades.

In relating lag time to risk, a frequent argument has been that more quantitatively and qualitatively advanced latent capacities are inherently more dangerous. Consider the following set of propositions:

1. The more quantitatively and qualitatively developed a state's latent capacity, the lower the direct financial and technical costs of "going nuclear," and the shorter the lag time.
2. The lower the direct costs of going nuclear, the easier it is to make the decision to do so, all else being equal.
3. The shorter the lag time, the greater the perceived applicability of the "nuclear option" to events at hand, and hence the greater the likelihood of making a proliferation decision.

4. The shorter the lag time, the less likely are possibilities for timely and effective antiproliferation intervention and treatment.

The first proposition stands without argument. Clearly, a prospective nth country will realize savings in the direct costs of a nuclear weapons program if it can divert or convert resources from an existing civil nuclear infrastructure. (This ignores any costs incurred by the civil nuclear program.) Perhaps more important than financial costs, direct resource demands are reduced considerably as a country moves from a no infrastructure latent capacity to a moderate infrastructure, then to advanced infrastructure latent capacity. In many ways the financial burden of a nuclear weapons program is the easiest of all to bear. Among the heaviest burdens would be the demand for skilled craftsmen, toolmakers, tradesmen, and industrial/construction materiel.

Proposition 2 places technological capability in the role of aid/constraint to nuclear proliferation, not cause. It suggests that relative technological capability may interact with nuclear propensity. Consider two countries in similar situations with comparable moderate to strong nuclear propensities. For one the financial and resource demands are small compared with its national resource base, while for the other the demands are fairly high. Proposition 2 argues that the country facing the smaller relative burden will be more likely to make a proliferation decision, all else being equal.

The analyses presented in the preceding chapters showed no evidence of such catalytic effects. However, keep in mind that for the most part the data base was dominated by several dozen of the most industrially advanced states in the world. For those countries the industrial-related resources (e.g., tradesmen, industrial and construction materiel) were never in short supply. As a matter of fact, in most cases these industrial-type resource demands were satisfied well before the nuclear age. However, the countries just acquiring latent capabilities possess significantly different technical profiles. Only recently have they acquired the rudiments of an industrializing economy. Countries not expected to acquire latent capacities until the mid-to-late 1980s lag even further in this regard. This suggests that, in the future, relative nuclear infrastructure may indeed have a catalytic effect on proliferation decision making—something that has not yet manifested itself.

Yet one also must consider the underlying reason for interest in the nuclear weapons program. On the one hand, in situations where decision makers perceive a nuclear weapons program as "lifesaving"—analagous to cancer treatment—then interest is likely to be fairly insensitive to cost within reachable boundaries. On the other hand, in situations where a nuclear weapons program is perceived more as cosmetic surgery—like a nose job—interest is apt to be more sensitive to cost differentials. This latter point suggests that any assessment of risk associated with relative nuclear weapons program resource demands must be explicitly tied to case-by-case evaluations of the strength of decision makers' underlying motivation for "going nuclear." So the converse may be true also: nuclear propensity may affect the way decision makers perceive the resource burdens of a nuclear weapons program.

Proposition 3—shorter lag times make proliferation decisions more likely—is an often-voiced hypothesis. Intuitively, it makes sense. If a situation arises where decision makers believe nuclear weapons would be useful, a short (or timely) lag time would argue for a proliferation decision. Conversely, a lag time of many years would imply little direct utility to matters at hand (Greenwood, Feiveson, and Taylor 1977, 150-51). This is, however, a wholly theoretical argument for which there is scant historical evidence. Moreover, exactly the reverse argument can be made—and for this there is some historical support. Long lag times may provide incentives for explicit decisions to shorten existing lag times as decision makers chose not to get caught "short" in the future. "Nuclear option building" is one possible outcome. Conversely, a short lag time may provide a buffer against proliferation decisions that result in actual weapons. Since with short lag times nuclear weapons could be produced in "fairly short order," there is no need for an immediate decision. In a sense, short lag times allow decision makers to put off the critical decision to move to nuclear weapons fabrication and production. This was certainly the case in India during the 1960s, in Sweden, in Canada, and to some extent in the Federal Republic of Germany. My only point here is to argue that short lag times do not necessarily imply a greater likelihood of proliferation decisions.

Proposition 4 is the most significant of the technical arguments. Shorter lag times will reduce possibilities for timely and effective antiproliferation intervention—given there is interest in undertaking this kind of activity. Lag time is, in fact, a surrogate measure for the amount of time available for direct antiproliferation intervention, assuming that United States intelligence is able to detect the proliferation decision within a very short time after it has been made. It is more realistic to assume, however, that intelligence will not promptly detect such decisions and that detection will therefore follow from the observation of postdecision indicators of "suspicious" behavior. Thus in almost all cases lag time becomes the maximum limit of the sum of time to detect and intervention time. (It is doubtful that countries could reduce lag time to less than a month or so without having made a proliferation decision.) This risk factor is compounded by the fact that decreasing lag times are a function of increasingly advanced nuclear infrastructures. More advanced nuclear infrastructures, of course, imply fewer potential indicators of overt proliferation behavior. As a result, the problem of shorter detection times is compounded by a reduction in the number of observables relevant to detection. There may even be instances where no observables exist and where the only hope of detection comes from human intelligence sources (spies and informers).

In sum, lag time is an important risk factor because of the way it reflects the time constraints within which an antiproliferation effort must be initiated if a proliferation decision is to be prevented or reversed. It is an indicator of how quickly a country can move from nonnuclear weapons status to nuclear weapons status given the desire to do so.

For the sake of simplicity, lag time is scaled as a three-level ordinal variable: short, medium, long. These levels are tied to nuclear infrastructure capabilities described earlier, where a short lag time connotes a length of about one year, a medium lag time is about four years, and a long lag time is about six years. A fourth level might be boundary lag time, connoting about ten years or more as time to bomb. To forecast lag time into the more distant future, industrial projections and technical assessments of growth plans can be used to estimate values for the technical indicators described in chapter 2. Latent capacities and nuclear infrastructure levels can then be forecast.

FORECASTING FOR THE NEAR TERM

To illustrate this forecasting concept, let us examine the near term population at risk. This exercise can be thought of as a mock briefing to the president, who has suddenly become interested in the status of nuclear proliferation. "How do things look?" he asks.

Lag time and nuclear propensity values were recorded as of mid-1982. Salience, from survey responses, is for the period 1982-87. Because I do not want to presume administration nonproliferation policy, treatability has not been incorporated as a direct measure into the risk analysis presented here. Instead, the particular (in)sensitivities of interesting cases will be noted in the course of the text. The data are given in table 16. For each country, lag time, nuclear propensity, and likelihood levels are shown. Several general obser-

TABLE 16
RISK FACTOR VALUES IN THE NEAR TERM

	Risk Factor		
Country	Nuclear Propensity[b]	Salience[b]	Lag Time[c]
Japan	3	1	1
Federal Republic of Germany	3	1	1
South Africa[a]	1	2	1
Belgium	3	3	1
Italy	3	2.5	1
Netherlands	3	3	1
Canada	3	3	1
Argentina	3	2	1
Spain	3	2.5	1
Israel[a]	1	1	1
Pakistan[a]	1	1	2
German Democratic Republic	3	2	2
Czechoslovakia	3	2.5	2
Switzerland	3	2.5	2
Norway	3	3	2
Brazil	3	2	2
Taiwan[a]	2	2	2
Yugoslavia	3	1.5	2
Sweden	3	3	2
South Korea[a]	2	1.5	2
Austria	3	3	2
Australia	3	3	3
Egypt	2	1	3
Rumania	3	2.5	3
Turkey	3	1.5	3
Greece	3	1.5	3
Iran	1	1	3+
Libya	1	1	3+
Iraq	1	1	3+
Mexico	3	2	3+
Chile	3	2	3+
North Korea	3	1	3+
Cuba	3	1	3+
Nigeria	2	1.5	3+
Algeria	1	2	3+
Finland	3	3	3+

[a]Has made previous proliferation decision.
[b]1 = high; 2 = moderate; 3 = weak.
[c]1 = short lag time; 2 = moderate lag time; 3 = long lag time; 3+ = boundary lag time.

vations are in order. From table 16 we can see that only six of thirty-six cases have high nuclear propensities, and only five have moderate nuclear propensities. Thus, in terms of likelihood, about 30 percent of these countries seem to have some interest in acquiring nuclear weapons (17 percent for sure). In terms of salience, 28 percent have high salience and another 33 percent have moderate salience, so 61 percent of these countries represent something to worry about in the consequences of their going nuclear. Last, 28 percent of the cases have short lag times and about 30 percent have moderate lag times. Potentially, 58 percent—twenty-one countries—could be producing nuclear weapons within four to five years. But the most useful information is found in the joint distributions of the data.

Table 17 presents the population at risk currently characterized by short lag times. These are countries that, all else being equal, could produce nuclear weapons within a median time of a year or so after making a proliferation decision. From an antiproliferation perspective, detection times of proliferation activity will be short, and detection itself will be difficult (except for those states pursuing the *dedicated facilities route*). In any case, once detection had occurred, little time will remain for antiproliferation intervention.

Looking across the first row, Israel appears in cell (111) and South Africa and Argentina in cell (112). (The order is lag time, nuclear propensity, and salience: 1 connotes *short* lag time, or high values of nuclear propensity/salience; 2 connotes *moderate* values; 3 connotes *long* lag time, or *weak* values of nuclear propensity/salience.)

TABLE 17
JOINT DISTRIBUTION OF THE POPULATION AT RISK: LAG TIME SHORT (1)

Nuclear Propensity	Salience		
	High (1)	Moderate (2)	Weak (3)
High (1)	Israel[a]	South Africa[a] Argentina	
Moderate (2)			
Weak (3)	Japan Federal Republic of Germany		Belgium Italy Netherlands Canada Spain

[a]Believed to have made a proliferation decision already.

Since neither Israel nor South Africa is recognized as having tested nuclear weapons, they continue to be treated by some as near-nuclear countries. They are included in the population at risk table for completeness, even though I assume that their governments have already made proliferation decisions.

Israel's salience is higher than that of South Africa, owing to the greater effects that Israeli proliferation would have on politico-military stability in the Middle East and on regional incentives for further nuclear proliferation, compared with the likely effects that South African proliferation would have in Africa. In the near term, these two countries are among the greatest proliferation risks.

Argentina is an important case that illustrates the dynamism of nuclear propensity. Until April 1982, Argentina's nuclear propensity was low. The Falkland Islands dispute, which could not have been forecast, radically altered Argentina's motivational profile. Argentina's military was humiliated, and the credibility of its alliance with the United States (Rio Treaty) was called into question when the United States sided with Britain. The result was an ominous shift in Argentina's nuclear propensity to the high category. Thus, during 1982 Argentina jumped from cell (132) to cell (112). In this instance a proliferation decision is to be expected. Because of Argentina's short lag time and its moderate salience, the risk analysis suggests that beyond Israel and South Africa (which probably already have nuclear weapons) Argentina is now the most significant case in the population at risk. Argentina's near-term treatability is probably low. Strained relations between the United States and Argentina during the 1970s (over human rights and nuclear energy issues) made Argentina turn increasingly toward Europe for economic and technical assistance and for trade. Self-imposed restrictions on arms sales by the United States also led Argentina to establish strong ties with other (European) arms suppliers. Thus little basic economic, military, or political leverage is left. Effective "treatment" would require close collaboration between most, if not all, major suppliers of nuclear technology, military technology, and industrial/economic aid. Of course this would necessarily require Soviet acquiescence.

Moving down to the third row—corresponding to weak nuclear propensity—we find the Federal Republic of Germany and Japan in the high salience category and Belgium, Italy, the Netherlands, Canada, and Spain in the weak salience category. None of these countries appear on the verge of making proliferation decisions. However, the potential dynamism of nuclear propensity coupled to the short lag times of these states must be kept in mind. In cell (131) we have the Federal Republic of Germany and Japan. Their high salience makes it especially important to ensure that their nuclear propensities remain low. Fortunately, domestic and international conditions are such that there is a strong array of disincentives. Moreover, these two countries are extremely susceptible to United States diplomatic and economic actions. Special political considerations—resulting from their roles in World War II—make the Federal Republic of Germany and Japan highly treatable.

In cell (133) are Belgium, Italy, the Netherlands, Canada, and Spain. All have very stable low nuclear propensities, and Spain's prospective entry into NATO should further reduce its nuclear propensity. In all these cases the arrays of incentives and disincentives make any shifts to higher nuclear propensity categories very unlikely in the near term. Their weak nuclear propensities and salience leave their short lag times as the only important contribution to proliferation risk.

Table 18 contains the countries with moderate lag times. Here there is "breathing room" of several years between a proliferation decision and the initial output of nuclear weapons. Thus these countries would not *at present* command the same urgency for intervention and treatment as do countries at the same nuclear propensity/salience levels but with short lag times. Keep in mind, however, that this is a temporary situation for many of these countries. As their nuclear infrastructures continue to develop, many will make the transition to short lag times. It follows that the

TABLE 18

JOINT DISTRIBUTION OF THE POPULATION AT RISK: LAG TIME
MODERATE (2)

Nuclear Propensity	Salience		
	High (1)	Moderate (2)	Weak (3)
High (1)	Pakistan[a]		
Moderate (2)		Taiwan South Korea	
Weak (3)		German Democratic Republic Brazil Yugoslavia	Switzerland Czechoslovakia Norway Sweden Austria

[a]Believed to have made a proliferation decision already.

importance of treating countries with moderate lag times now lies in the fact that there is time to implement an orderly and coherent set of programs to affect their long-term nuclear propensities.

The only country currently in the high nuclear propensity row is Pakistan—cell (211). Without further intervention and treatment, Pakistan will become a nuclear power within the next several years. Pakistan's high salience makes intervention even more critical. Not only would a Pakistani nuclear stockpile inflame military tensions with neighboring India, but Pakistan's cultivation of a role for itself within the Islamic community raises implications for Persian Gulf and Middle East stability as well. Pakistan's treatability is difficult to assess. In part there is a problem of "the tail wagging the dog" here. Pakistan's Arab benefactors have given it a measure of political, economic, and military insulation from the United States. There have been severe political strains between the United States and Pakistan for more than half a decade. Now, as a result of the Soviet occupation of Afghanistan and the fall of the shah of Iran, Pakistan has regained its strategic importance to the United States. There may well be an asymmetry of "need"—the United States needs Pakistan more than Pakistan needs the United States. In short, the United States may have more leverage with Pakistan than it would dare use under current conditions. Pakistan probably should be grouped as having low treatability.

Moving down one row to the moderate nuclear propensity level, we find both Taiwan and South Korea in cell (222). Previous analysis reveals that both these countries had strong nuclear propensities during the mid-1970s that converged with technical capability. Case study evidence for corresponding proliferation decisions does exist. However, timely intervention by the United States, involving highly tailored treatments, produced decision reversals. Subsequently, Taiwan's and South Korea's nuclear propensities have dropped to the moderate level. This suggests that government interest in nuclear option building continues within these countries. Moreover, the mixes of incentives and disincentives arrayed against them are fragile and could easily give way. A loss of United States alliance credibility, for example, could cause their nuclear propensities to shoot up to the high category. As with the Federal Republic of Germany and Japan, however, the United States possesses considerable political, military, and *economic* leverage over these countries that suggests they would be particularly responsive to antiproliferation intervention and treatment.

In the weak nuclear propensity row we find the German Democratic Republic, Brazil, and Yugoslavia in cell (232) and Switzerland, Czechoslovakia, Norway, Sweden, and Austria in cell (233). The former group has moderate salience, and the latter has weak salience. The most interesting case here is Brazil. Though its nuclear propensity is currently weak and stable, Argentina's acquisition of nuclear weapons could push Brazil's nuclear propensity up considerably higher. Brazil is also quite likely to make the transition from moderate lag time to short lag time during the next several years.

Long lag time countries and boundary lag time countries (denoted with a plus sign) are listed in table 19. Since their times to bomb are

TABLE 19
JOINT DISTRIBUTION OF THE POPULATION AT RISK: LAG TIME LONG (3)

Nuclear Propensity	Salience		
	High (1)	Moderate (2)	Weak (3)
High (1)	Iran+[a] Libya+ Iraq+	Algeria+	
Moderate (2)	Egypt	Nigeria+	
Weak (3)	North Korea+ Cuba+	Turkey Greece Mexico+ Chile+	Australia Rumania Finland

[a]Plus sign means boundary lag time.

of the order of a half-dozen years or more, a nearsighted United States administration can look upon them as someone else's problem. There is little reason for time-urgent intervention and treatment. But, to reiterate, previous analysis shows that for many states the arrays of incentives and disincentives producing high nuclear propensities are most enduring. Cases such as Israel, South Africa, and China demonstrate that when technical capability converges with long-standing high nuclear propensity, proliferation decisions result. In essence, persistently high nuclear propensities represent "accidents" waiting for the opportunity (i.e., technical capability) to happen. It is interesting that the outcome of convergence appears to be independent of the dynamics between the factors: whether motivation converges with technical capability, technical capability converges with motivation, or they converge simultaneously, it produces the same effect.

As can be seen in the first row of table 19, Iran, Algeria, Iraq, and Libya all now have high nuclear propensities. Iran, Iraq, and Libya have high salience—cell (311)—while Algeria has moderate salience—cell (312). (Note that all three actually have boundary-condition lag times and thus technically should be in a fourth level lag time cell: [411] and [421]).

The fall of the shah radically altered Iran's nuclear profile. The good news is that it brought to a halt Iran's ambitious plans for civil nuclear development—plans that might have left Iran with an advanced nuclear infrastructure within a decade or so. The bad news is that the undoing of Iran's alliance with the United States removed a significant disincentive against nuclear proliferation. Iran's treatability declined radically. Most recently, the Iranian government has announced plans to restart the civil nuclear program outlined during the shah's rule.

In Libya, Qaddafi's several efforts to purchase nuclear weaponry on "the international market" reflect a more entrepreneurial dimension of high nuclear propensities. His simultaneous moves to finance the Pakistani nuclear program in the hope of skimming expertise and weapons stand as further testimony to his creativity. That these endeavors ended in failure should not be allowed to obscure the fact that they represent the more fundamental drive to "go nuclear." In time, perhaps as the result of an indigenous civil nuclear program through Soviet aid, Libya could acquire a latent capacity. The bizzare behavior of both the Iranian and the Libyan regimes, coupled with their high salience, suggests that they should be treated as very special cases within the population at risk—perhaps placed ahead of states with shorter lag times for intervention and treatment. (They may require long periods of "therapy.") Iran's treatability is low, if it exists at all. The Libyan case is difficult to judge. Exactly how much influence do the Soviets have with Qaddafi, and how much are they willing to jeopardize their position in North Africa in the name of nonproliferation? We can only assume that Libya has low treatability.

As I have already noted, Iraq's nuclear propensity and technical capabilities have been affected by several recent events. The destruction of Iraq's nuclear research facilities by Israeli aircraft has set back Iraq's nuclear research and development timetable by many years. Moreover, the French and the Italians have reconsidered some aspects of their original agreements with Iraq for the transfer of nuclear technologies. The Iran-Iraq war has drained the Iraqi treasury and severely cut revenues from oil exports, thus crippling Iraq's ability to import nuclear and industrial equipment and expertise. On the motivational side, one must presume that the aftershocks of Iraq's loss of the Iran-Iraq conflict (i.e., being pushed out of Iran), the humiliation of the Israeli attack on Osiraq, and

Iraq's regional power pretensions will keep its nuclear propensity strong. Given the current state of Iraqi-United States and Iraqi-Soviet relations, it is doubtful that Iraq is susceptible to treatment.

Jumping down to the second row—moderate nuclear propensity—we find Egypt with high salience and Nigeria with moderate salience. The nuclear propensity of Egypt is highly volatile (as a function of the array of incentives and disincentives) and could shift to the high category under a number of possible contingencies. Moreover, Egypt's incentives to go nuclear are multidimensional—that is, no single treatment can be expected to significantly alter its nuclear propensities. Nigeria's nuclear propensity is more stable and anchored more or less in a single issue: the South African conflict. Again there is a substantial amount of time to act in these cases, but we must keep in mind that these cases represent tomorrow's crises in incubation.

Finally, at the weak nuclear propensity level we find a diverse mix of cases. There are worrisome (from a long-term perspective) cases in the high salience category such as North Korea and Cuba and a conflict pair in the moderate salience category: Turkey and Greece. The latter pair is noteworthy because of the symmetry in their risk profiles. This should make them easier to treat. (Contrast this with, say, Pakistan and India or Iraq and Israel, where technical "catch-up" may be an independent incentive in its own right.) Yet if they are left untreated, a nuclear crisis could well erupt in the Aegean in the early 1990s.

IMPLICATIONS OF THE POPULATION AT RISK

This brief exercise in forecasting for the near term offers some insights into generic approaches to nonproliferation. To draw them out, it is useful to group the population at risk in several basic classes of proliferants: the hard core, the middle core, and the soft core.

Hard-Core Proliferants

These are countries that already have made decisions to become nuclear weapons powers. The intensity of their motivational profiles is such that, no matter how many technology barriers are thrown up in their paths, they will continue to pursue a nuclear weapons capability. They are characterized by high (and stable) nuclear propensities, low treatability, and high salience. Examples include Pakistan, Iraq, and Libya.

The existence of hard-core proliferants testifies to both the importance and the insufficiency of a nonproliferation effort based largely on technological control. Technological controls raise the ante for entrance into the nuclear club. To enter, a country must be willing to pay a special price. But of course this is precisely the definition of hard-core proliferants. For the hard core, nuclear weapons are worth the price. What technological controls do in such instances is buy time. Without any technological controls countries such as Libya could very well acquire (import) advanced latent capacities within a half-dozen years. By purchasing turnkey nuclear facilities, they could focus domestic, scientific, and technical talent on the weapons design work. The quality and production capacity of the fissile material production facilities would probably exceed by far what they could have accomplished indigenously over much greater periods.

However, technological controls, more stringent safeguards, or even the collapse of the international civil nuclear technology market will not bring the hard core under control. Eventually—a decade or more from now—even the states with boundary lag times will acquire the wherewithal to proceed with nuclear weapons production by indigenous (dedicated) means. The production rate might not be large, but it would certainly be of sufficient dimensions to threaten regional military and political balances. A move away from international sales of nuclear technology by the major suppliers would only lead to more *indigenous* nuclear power research programs. Numerous research and laboratory-scale fissile material production facilities would spring up. The modicum of control, monitoring, and safeguards that currently envelop international nuclear technology sales would disintegrate. Indigenously developed research facilities and laboratories—spawned by cutoffs in nuclear technology commerce—might come to pose greater proliferation risks than would the lesser number of commercial-scale facilities likely to be in demand.

In any case, technology controls do buy time—but time for what? The answer will come as no surprise: to significantly alter motivational profiles. Lowering the nuclear propensities of the hard-core proliferants will not be easy. After all, that is why they are labeled hard core. The difficulty is that altering these states' motivational profiles implies resolving some fundamental problems in world politics. Cease-fires, truces, and armistice agreements are not sufficient. The need here is for the kind of political breakthroughs typified by the Israeli-Egyptian Camp David Accords.

Unless efforts are made to alter the motivational profiles of the hard core, technological controls will have little long-term value to nonproliferation efforts (other than to make these countries someone else's problems). Moreover, the net effect will be to bunch all the hard core countries at the far end in time—they will proliferate in clusters within a short time of one another owing to the retardation of their acquiring latent capacities. This could be an even more threatening situation than if they had been "allowed" to proliferate in a more natural order. Again, this is a danger if policymakers do not take advantage of the time bought through technological controls.

Middle-Core Proliferants

These are countries for which there exist political/military/economic costs beyond which going nuclear is not worth the price. Their nuclear propensities may be moderate or high, but so are their treatability levels. Middle-core proliferants have other priorities of equal or greater weight than those associated with their motivational profiles. They differ from the hard core in that they are not willing to "pay the price" of nuclear proliferation in the face of direct nonproliferation intervention and treatment.

Here technological barriers, safeguards, and other international agreements are useful in containing nuclear ambitions. Political and economic relations also play a crucial role, as in the cases of South Korea and Taiwan.

Work on motivations is important, if for no other reason than to ensure that fewer of these states move into the hard core. Argentina, for example, was probably among the middle core before the Falkland Islands dispute erupted but may have since moved much closer to the hard-core group.

Soft-Core Proliferants

Here one finds countries with low nuclear propensities and high treatability. They are uninterested in nuclear weaponry. To these countries technological controls are largely an inconvenience (and, for them, unnecessary), but they are willing to put up with them to ensure that the middle-core and hard-core proliferants are held in check. Indeed, technology controls may be one reason these countries can afford to be soft core.

A FUTURE UNLIKE THE PAST?

The forecasting exercise presented above assumed that the nuclear proliferation process in the future would be analogous to that observed in the past (Nacht 1981; Quester 1981). Modes of decision making, the relevance and effects of various motivational factors, geopolitical considerations, military-technical cost/benefit calculations, and the politics and structure of the prevailing nonproliferation "regime"—all important in determining the rate, pace, and scope of the nuclear proliferation process—were treated as fairly consistent over time. But are we justified in making such assumptions? And if so, what are the implications?

For example, it has been pointed out that a substantial amount of the data base is composed of European industrial states. Yet it is widely believed that much of the nuclear proliferation related activity in the coming years will take place in developing countries in the Third World. We also must acknowledge that the political and economic character of the international system in the 1980s and 1990s is likely to be quite different from that of the 1950s and 1960s. While such points initially seem to suggest a weakness in the forecasting method, in fact they underscore its inherent strength.

Policymakers need some basis for evaluating events and making decisions. Their own perceptions of events and the advice they receive from others are grounded both in interpretations of past events (history) and in impressions of how things have changed. In this respect the historical data base used here represents an empirical baseline for understanding proliferation behavior. Should the future be unlike the past, we will need such a baseline to gauge how the future is different and what the corresponding implications are. For instance, previous analysis suggests that domestic turmoil adds little to the nuclear propensity of states. If one believes this factor is likely to prove more relevant in the future, then estimating its increased effects is made easier by knowing its prior effects. The forecasting method incorporates both the baseline data and procedures for blending in new information. If the strength of a particular motivational variable is believed to have increased, this can be také into account by manipulating the numerator and denominator in the nuclear propensity calculations. Here the empirical value found in table 13 can be considered the Bayesian prior, and integers can be added (or subtracted) from the appropriate expression so as to conform to impressions of change (Winkler 1972; Schmitt 1969). In this way several "alternative futures" can be investigated in a systematic fashion. Each will begin with the common historical prior information and will vary according to different assumptions about the future. Of course variables can be totally deleted if they are deemed irrelevant, while new variables can be added.

It follows that the forecasting exercise in this chapter should be considered a baseline risk analysis for a future that plays by much of the same rules as the past. Other forecasting exercises should be undertaken under different assumptions and compared with that baseline.

Thus, in contrast to the popular view of forecasting as crystal ball gazing, the notion of forecasting developed here does not emphasize predicting the future. Instead, it provides an analytic structure in which data can be assembled, uncertainties can be highlighted, opportunities can be noted, and risks can be evaluated.

Concluding Observations

At the heart of this study is the contention between two very different notions regarding the fundamental cause of the proliferation of nuclear weapons states. Must a country that is capable of manufacturing nuclear weapons inevitably do so, or—aside from providing opportunity—are proliferation capability and proliferation outcome independent? Despite numerous efforts, the data revealed no support for a technological imperative: one cannot distinguish between countries that go nuclear and those that do not merely by examining relative capabilities. Yet a systematic analysis of the historical evidence does demonstrate strong and unambiguous support for the motivational hypothesis.

Technology is, of course, one of the two necessary ingredients of the nuclear proliferation process. Technical controls, nuclear safeguards, and limitations on the transfer of certain critical technologies can "buy time" by slowing the pace and scope of nuclear proliferation (Dunn 1982; Potter 1982). But trying to prevent the emergence of new nuclear weapons countries solely through the control of technical means amounts to treating the symptoms while ignoring the disease. Over the long run, the "cure" for the problem of nuclear proliferation lies ultimately with the reduction of national nuclear propensities. From this perspective, the diversion of fissile materials from civil nuclear power installations will become less of a proliferation threat as the motivation for proliferation is reduced.

The findings of this study show that *many* distinct motivational profiles (taken in conjunction with adequate technical ability) can produce nuclear proliferation events. Conversely, they also show that proliferation decisions are reversible when motivational profiles change to produce lower nuclear propensities. It seems that, for the long term, the most effective antiproliferation strategies will be those that are most "in tune" with the process of nuclear proliferation and, correspondingly, that can devise and implement narrowly tailored corrective measures—that is, tailored to the prospective proliferant's individual motivational situation. This will require early warning of impending proliferation risk and an understanding of the factors and dynamics that are at play.

Appendix A
Historical Decisions to Initiate Nuclear Weapons Programs

Though there are but six nations known to possess operational nuclear weapons—and one additional suspect—there have been at least thirteen historical decisions to initiate nuclear weapons programs. These cases denote only instances where a latent capacity already existed and the decision involved transforming it into an operational capability.

The years attributed to the various decisions are based on a survey of historical materials. The expected error of plus or minus one year is insignificant for my purposes, given that most of the putative influences considered in this study tend to persist over a several-year time span.

Nazi Germany (1940) and Imperial Japan (1941)

The theoretical possibility of devising an explosive based on nuclear fission was recognized by scientists of many nations even before the beginning of the Second World War. Thus, there should be little surprise that scientists in both Nazi Germany and Imperial Japan pursued such research during the war. The historical record reveals that their military authorities began to evince more than casual interest in atomic explosives during 1939 to 1942.[1] Explicit research programs were summarily undertaken. The German War Office, for example, began active funding of nuclear research in 1940, simultaneously taking control of all research facilities and stocks of nuclear-related material (e.g., uranium, heavy water). The Japanese government intervened on a more direct scale somewhat later, in 1941.

The United States and the United Kingdom: The Manhattan Project (1942)

The first successful atomic bomb program was, of course, the Manhattan Project. This joint American-British-Canadian undertaking (with Canadian participation under British auspices) eventually ended all doubt regarding the scientific feasibility of creating fission chain reactions. Here the documentation is clear that sometime between June and September 1942 presidential authorization was given for a full-scale program (Brown and McDonald 1977; Groves 1962; Lawrence 1959; Gowing 1964). Furthermore, budgetary data released by the United States Atomic Energy Commission (1956, 20) revealed that initial funding for the Manhattan Project began 1 July 1942. Therefore 1942 was coded as the decision date for the United States and for the *first* British program.

The Second British Program (1947)

Recorded above was the first British decision, resulting in a program carried on in collaboration with the United States (1942). However, the United States decision to "keep the atomic secret" to itself at the conclusion of the Second World War meant that if the British wanted atomic weapons they had to strike out on their own. The British did indeed make an independent decision to pursue the elusive fission bomb, and several well-researched studies point to (January) 1947 as the appropriate decision date (Gowing 1974, vols. 1 and 2; Groom 1974; Pierre 1972).

The Soviet Program (1942)

Detailed information about the circumstances surrounding the Soviet program is scant, to say the least. There is substantial evidence of Soviet awareness and interest in fission explosives between 1939 and 1945 (Kramish 1959; Modelski 1959; Holloway 1980). But the war with Germany interfered with substantial research. The initial German steamroller forced the Russians to evacuate their industry over the Urals. Scientists were put to work on the most pressing and near-term problems. Do not forget, the "atomic bomb" was then still just a theory. Thus it was not until the Soviets halted the German advance that they could consider instituting an atomic weapons program. Undoubtedly, intelligence provided them with information on what their allies were doing in the American Southwest.

A number of Soviet writings on the subject point to the summer of 1942 as the time when Stalin and the State Defense Committee formally approved the initial Soviet atomic bomb program.[2] However, the Soviet program did not critically get under way until the fall of 1943. And it did not acquire the high priority of the United States-United Kingdom Manhattan Project until the summer of 1945.

The French Program (1956)

Dating the French "decision" to undertake the manufacture of nuclear weapons was complicated not by a lack of information, but by a surfeit. Depending on one's notion of what constitutes a "final government decision," various scholars have suggested 1955, early 1956, and late 1956 as the date of the final French decision.[3] The question of what exactly constitutes a decision within the realm of French politics is at the root of the problem.

Two pieces of information seem particularly relevant. In March 1956 Premier Faure stated that France would eventually acquire nuclear weapons (Dhar 1957, 77, 197). Then in November 1956 Faure declared that France would indeed develop nuclear weapons and had been conducting the appropriate research *for months*. Since nuclear research in France went back several decades, Faure's reference to months could only have pertained to explicit weapons research. Thus, using the French premier's statements as a guide, I coded 1956 as the decision date for the French nuclear weapons program.

The Chinese Program (1957)

The Chinese, like their Soviet colleagues, have not been forthcoming with details about their early program experiences. Nevertheless pieces of information are available that make an estimate of the date of the program start-up decision possible.

According to the Soviets, the Chinese Atomic Energy Committee (under the Chinese Academy of Science) was established as early as 1953 (Gorbacher 1981; Akimov and Panior 1973). By 1955 there were thirty-six nuclear research laboratories, including a uranium mining and milling operation in Sinkiang province (Gorbacher 1981; Wu 1970). In 1957 the Chinese Institute for Atomic Energy was set up, and within a year China's first research reactor was operational (Halperin 1965, 71).

Also occurring in 1957 was a secret Chinese-Soviet agreement on nuclear sharing. The Soviets reportedly agreed to provide the

Chinese with a working atomic bomb along with some technical assistance (Hsieh 1964b, 106; Roberts 1970, 66). Though the agreement was subsequently nullified by Moscow, over the several years it was in force the Soviets did provide considerable training and assistance in general nuclear science and engineering.

The Chinese program to manufacture nuclear weapons actively got under way sometime in 1958 (Horner 1973, 235; Chang 1971, 209; Hsieh 1964a). The decision to go ahead clearly came immediately prior to this. The Soviets claim that Mao, in a secret meeting of Chinese military and scientific experts, proclaimed that China must have "the atom bomb. We cannot do without it" (Gorbacher 1981). In this respect the subsequent Sino-Soviet nuclear technology agreement may have been important to Chinese program planning. Thus it seems that sometime between 1956 and 1958 formal authorization was given to begin. Because of the uncertainty here, 1957 is assumed to be the date of the Chinese decision to proceed with nuclear weapons production.

The Indian Case (1965)

Since the early 1950s, India has had a vigorous nuclear research program (Kapur 1976; Paloose 1978; Meyer 1981b). From its inception and until 1966 that program was under the personal direction of Dr. Homi Bhabha. Bhabha held every top-level scientific and government position connected with nuclear energy save one: minister of atomic energy, which was held by the prime minister. From the late 1950s on Bhabha agitated for Indian research on peaceful nuclear explosives (PNEs). Prime Minister Nehru resisted this line of nuclear development, though he allowed Bhabha to direct India's nuclear program along lines that would always permit an easy transition to nuclear explosives development.

On the death of Nehru, Shastri became prime minister. Though initially inclined, like Nehru, against nuclear explosives development, in late 1965 Shastri approved Bhabha's Subterranean Nuclear Explosives Project (SNEP). Within two to three years India would be ready to test its nuclear device. When Shastri and Bhabha died in January 1966, SNEP was canceled by India's new prime minister, Indira Gandhi.

The Israeli "Program" (1968)

Clearly, the Israeli case involved a "judgment call." Unlike the recognized nuclear weapons countries—and quasi-nuclear India—

Israel has yet to demonstrate its operational capability with a test detonation. But one should not confuse a historical convenience with a technical necessity.[4] In this particular instance there was enough convergence between independent reports of an existing Israeli operational capability to warrant its inclusion here.

First, there were the leaks of United States intelligence reports about Israeli nuclear research (Dowty 1972; Haselkorn 1974; Harkavy 1977). Then there were rumored reports of functional Israeli atomic weapons during the 1973 Middle East War (*Time* 1976, 39-40). In particlar, one report described the Israeli government's rejecting an atomic weapons program in 1967, only to reverse its decision in 1968 (*Time* 1976, 39-40). Later there were several substantiated reports of the covert diversion of more than two hundred metric tons of uranium oxide to Israel in late 1969 (*Time* 1977, 31-34). Given the size of the Israeli reactor program and the availability of uranium (for research needs), this last piece of information certainly seems to support the proposition that a plutonium production (nuclear weapons) program was under way. Finally, a number of assessments by United States government officials, as well as foreign sources, suggested an Israeli operational capability emerging sometime in 1971-72.[5] Assuming that they required roughly four years from start to finish (see Chap. 2), this would place the Israeli decision in 1967-68.

The Indian Case (1972)

Though nominally labeled a peaceful nuclear explosion, the Indian test of May 1974 could equally well be considered an atomic bomb test. Clearly the Indian government was sensitive to the foreign policy implications of its test, and it certainly recognized its impact on Pakistan and the People's Republic of China. Thus, for all intents and purposes, the decision to build and test the device was assumed to be analogous to that of a "declared" atomic weapons program.[6]

Studies of the Indian nuclear explosives program suggest that engineering feasibility studies were initiated sometime between 1970 and 1972 (Seshagiri 1975, 13-14). But it was not unitl sometime between December 1971 and early 1972 that the decision was made to construct the device and conduct a test.[7] Accordingly, 1972 (the best estimate) was set as the date of the Indian decision to launch its program.

Appendix A

The Korean Case (1972)

In 1970 President Park set up two ad hoc working groups to study ways South Korea could improve its domestic arms industry.[8] One of these—the Weapons Exploitation Committee (WEC)—examined the possibilities of nuclear weapons production. A former WEC member, testifying as part of the "Korea Gate" scandal, revealed that in the early 1970s the WEC had unaminously voted for nuclear weapons production. President Park reportedly approved the recommendation.

The WEC recommendation also included the suggestion that a reprocessing plant be obtained overseas. (Apparently to be used in conjunction with Korea's power reactors to produce plutonium by the late 1970s to early 1980s.) In 1972 WEC members visited Israel, France, Norway, and Switzerland. A deal with France for a reprocessing facility followed.

The decision to proceed with the WEC recommendation appears to have been made in late 1971 or early 1972. Taking into account the Korean activities to investigate reprocessing options in mid-1972 and assuming that these activities followed on the heels of Park's authorization, I coded 1972 as the decision date.

The South African Program (1975)

In the late 1960s some South African political leaders raised the possibility of South African interest in peaceful nuclear explosives. While it is possible that some feasibility research was conducted, there is no evidence that there was a government decision to proceed in a serious way.

In August 1977 it was revealed that Soviet reconnaissance satellites had detected what appeared to be preparations for a nuclear test in a South African desert. United States reconnaissance satellites corroborated the Soviet data, and a flurry of diplomatic activity followed. That South African test never took place.

Nonetheless, the preparations themselves imply that the test probably would have occurred sometime in late 1977 or early 1978. Since South Africa's capability to manufacture nuclear weapons is based on an advanced nuclear infrastructure (its uranium enrichment facility), one could expect a two-year lag between the decision to proceed with nuclear weapons production and the appearance of the first nuclear weapon. This would give 1975 as the decision date.

Appendix B
The Technical Model

As I noted in chapter 2, the purpose of the technical model is to establish a systematic procedure for assessing national technical and industrial capabilities that are relevant to the production of nuclear weapons. While it would be ideal to have a precise time series on the number of specialists in each country trained in various fields of nuclear science and engineering, the level of development of key areas in metallurgical, chemical, and electrical industries, the scope and size of national nuclear research programs, and the quality of science and engineering in each country, such a data base does not exist. To be sure there are individuals—both in and out of government—who do possess such information on one or more countries for some given time period. But for our needs this is not sufficient. We require a complete time series for all prospective proliferators covering the entire time span 1940-82.

An alternative is to try to develop a reasonable impression of a given country's basic capabilities through surrogate indicators. These indicators are not intended to measure particular capabilities on a one-for-one basis, though, of course they will be correlated with certain specific capabilities. Rather, all the indicators taken together are required to get a general picture of the resource base of a nation and evaluate its ability to support the base case (low-technology) nuclear weapons program. The reader should thus keep in mind that the technical model is predicated on the synergy of the component indicators, not on their individual validity as indicators of technological prowess. Particular emphasis is placed on what I have called the "Sears, Roebuck" and "Yellow Pages" factors.

To better understand the underlying structure of the technical model, the discussion below begins with a brief description of the component parts of the base case nuclear weapons program, then turns to the indicators used in the technical model.

For the purposes of this study, a nuclear weapons program is any national program implemented with the explicit goal of producing operational and transportable nuclear explosives. Moreover, such a program would nominally have a planned output capacity equivalent to roughly one nuclear explosive device a year over several years. Whether such a program incorporates a long-term output schedule, a test series, or development of a delivery capability is treated as incidental and as external to the program itself.

The "heart" of a nuclear weapon is its fissile core. Under the proper conditions, the nuclei of the atoms that make up the core material will eventually split into fragments when bombarded by atomic particles called neutrons. It is this process of nuclear fission that releases nuclear energy. Moreover, certain kinds of atoms also release additional neutrons as they undergo fission. When the number of neutrons released by nuclear fission equals or exceeds that needed to cause fission, then a chain reaction is possible. In this instance the fission of one atom results in the fission of one or more additional atoms, which in turn cause more atoms to split. In the design of nuclear weapons, the goal is to create the conditions needed for an uncontrolled chain reaction.

A Plutonium-Based Nuclear Weapons Program

Though a number of elements and isotopes can theoretically form the core of a nuclear weapon, to date only plutonium 239 (Pu-239) and uranium 235 (U-235) have actually been employed. In designs that use Pu-239, the core is surrounded by a chemical explosive. When the chemical explosive is detonated, the plutonium core is imploded—driven inward—radically increasing its density. At the very high densities attained more neutrons are released by fission than are needed to produce fission. (Before implosion, the density of the plutonium core is such that more neutrons are needed to cause fission than are released by fission. Thus no chain reaction can occur.)

Plutonium is not a naturally occurring element and therefore cannot be mined. Instead it must be artificially created by transmuting U-238. This process of converting U-238 into Pu-239 is accomplished in a nuclear reactor. Indeed, it is the design, construction, and operation of the various facilities needed to produce the Pu-239 that represent the bulk of the activities—and difficulties—in a plutonium-based atomic weapons program.

TABLE 20
STEPS IN PRODUCING PLUTONIUM 239 ATOMIC WEAPONS

1. Acquiring of uranium source material
   a. Mining of uranium-bearing ores
   b. Milling of ores to extract "raw" uranium compounds
   c. Design and construction of chemical processing (conversion) facility
   d. Chemical processing to convert "raw" uranium compounds into usable uranium compounds (e.g., UO_2, UF_4)
2. Fuel fabrication (for production reactor)[a]
   a. Design and construction of fuel fabrication facility
   b. Fabrication of fuel-usable metal, alloys, ceramics, etc.
   c. Manufacture of fuel component (e.g., "canning in aluminum")
3. Plutonium production
   a. Design and construction of production reactor
   b. Reactor operation (plutonium production)
4. Plutonium extraction
   a. Design and construction of reprocessing facility
   b. Reprocessing plant operation (extraction of plutonium compounds)
   c. Chemical conversion of plutonium compounds into plutonium metal
5. Weapon fabrication
   a. Manufacture of plutonium core (metal sphere)
   b. Design and manufacture of nonnuclear components (e.g., chemical explosive, detonator)
   c. Weapon assembly

[a]Assumes natural uranium fuel loading for production reactor.

Table 20 outlines the primary activities. The first step is to acquire uranium source material—uranium-bearing ores from which the uranium for the fuel for the production reactor is obtained. Directly following the mining stage, the ores are milled to separate the "raw" uranium minerals from the other element-bearing minerals. This milling process is comparatively "simple and straightforward," including such steps as crushing, grinding, scrubbing, and heating the uranium-bearing ores. Next these "raw" unranium compounds are processed into any one of a number of fuel-usable uranium compounds by a variety of chemical conversion techniques.[1] Usually this involves converting U_3O_8 into U-metal, UO_2, UC_2, or a uranium alloy.

The fabrication of fuel elements for the production reactor in part depends on the design of the reactor—whether it operates on natural uranium fuel or slightly enriched uranium fuel. This consideration, however, does not complicate the analysis, since there are a number of compelling reasons why a natural-uranium fueled reactor is the only logical choice for an atomic weapons program.

First and foremost, among conventional fission reactors the rate of plutonium production per unit of thermal power output is higher for natural uranium reactors than it is for reactors fueled with slightly enriched uranium. I have excluded the nonconventional breeder reactor from consideration here—that is, with respect to a single-purpose atomic weapons program. Of the four basic plutonium isotopes: Pu-239, Pu-240, Pu-241, and Pu-242, only Pu-239 is valuable as weapons material. It is therefore desirable to produce plutonium with as high a fraction of Pu-239 as feasible. Second, unlike reactors fueled with slightly enriched uranium, natural uranium reactors can be refueled "on power" without a corresponding loss in operating time. And third, the rate of production of so-called weapons-grade Pu-239 relative to that of other plutonium isotopes can be more accurately controlled. Thus, natural uranium reactors possess all the characteristics of an optimal plutonium production reactor. Natural uranium fuel is also simpler to produce than its slightly enriched counterpart. As I explained in chapter 2, a specific class of natural uranium fueled reactors represents the least demanding, least expensive reactor design.

The fabrication of fuel elements for the plutonium production reactor is a metallurgical process. If the fuel is to be U-metal, it is usually either "canned" in aluminum, stainless steel, or zirconium or fabricated into an alloy, then shaped according to the reactor design. UO_2 is also "canned."

There are two general classes of production reactors that use natural uranium fuel: graphite-moderated reactors and heavy water (D_2O) moderated reactors. Irrespective of the type of reactor employed, the plutonium production process is the same. The fuel elements are loaded into the reactor, and a controlled fission chain reaction is maintained. The neutrons freed during the reactor's operation can do one of several things: (1) they can produce new fission events, (2) they can escape from the reactor, (3) they can be captured by various reactor materials, or (4) they can be captured by the U-238 atoms in the fuel. It is this last possibility, the capture of a neutron by an atom of U-238, that ultimately results in the creation of an atom of Pu-239.

At periodic intervals the "spent" or "burned" fuel elements (now containing Pu-239) are removed from the reactor and are replaced by fresh fuel loadings. The spent fuel elements are then submerged in a holding tank (a pool of water) for a month or more to allow thermal and radioactive "cooling." This cooling period eases the problem of postirradiation handling during the reprocessing stage.

A plutonium reprocessing plant chemically separates the Pu-239 from the residue of the spent fuel. Its design and construction must allow for remote handling of materials, owing to the radiological toxicity of plutonium and nondecayed fission products.

The spent fuel is subjected to a number of chemical conversions that, in the end, result in a pure plutonium metal. This plutonium metal is then sent on to the weapons fabrication laboratory.

The final stage is, of course, weapon fabrication. Here a team of scientists, engineers, and technicians (e.g., a physicist, a metallurgist, an electronics engineer, and an explosives engineer) assemble the various nuclear and nonnuclear components. A plutonium sphere of subcritical density must be manufactured. An inwardly focused shell of shaped chemical explosives is fabricated and positioned around the outer surface of the plutonium sphere. Then the fusing and detonation system are wired in. Various other auxiliary components most likewise be assembled, including a tamper, a neutron initiator-trigger, and so forth. The entire device is then sealed.

A Uranium-Based Atomic Weapons Program

In nuclear weapons that utilize U-235, a somewhat different (and less demanding) design can be employed. Here two masses of U-235 are held at separate ends of a chamber. When the chemical explosive is detonated, the two masses slam together, thus producing a single mass of sufficient material and density to permit a chain reaction.

Unlike plutonium, U-235 is found in nature. Therefore no production reactor is necessary in a uranium-based atomic weapons program. Unfortunately (from the perspective of future nth countries), U-235 exists in relatively small quantities compared with its more abundant sister isotope, U-238. In fact, in one thousand kilograms (one metric ton, or one tonne) of natural uranium there are only seven kilograms of U-235, but roughly 993 kilograms of U-238. (I have intentionally ignored the existence of minute quantities of other uranium isotopes.) Consequently, before a uranium bomb can be fabricated, the U-235 must be separated from the U-238.

Because U-235 and U-238 are chemically the same element, uranium, their chemical behaviors are identical and therefore nondiscriminating. One cannot separate U-235 and U-238 by chemical procedures. Instead, one must use isotopic separation techniques that take advantage of their differences in mass or electromagnetic properties or both. Without doubt, it is the design, construction, and operation of these isotope separation facilities that pose the greatest technical hurdles for the prospective nth country. Such facilities are more complex and costly than the corresponding plutonium-production facilities discussed previously.

Table 21 outlines the primary activities of a uranium-based program. Not surprisingly, the mining and milling operations are identical to those found in a plutonium-based program. Once the ore has been milled, it is similarly shipped on to a chemical processing plant.

TABLE 21
STEPS IN PRODUCING URANIUM 235 ATOMIC WEAPONS

1. Acquiring of uranium
   a. Mining of uranium-bearing ores
   b. Milling to extract uranium compounds
2. Uranium compound conversion
   a. Design and construction of conversion facility
   b. Chemical conversion of uranium compounds into UF_6
3. Isotope separation (enrichment)
   a. Enrichment facility design and construction
   b. Enrichment facility operation (separation of U-235 and U-238)
4. Chemical reconversion to uranium metal
   a.   Conversion of UF_6 compound consisting of U-235 into U-235 metal
5. Weapon fabrication
   a. Manufacture of uranium core (metal hemispheres)
   b.   Design and manufacture of nonnuclear components (e.g., chemical explosives, detonator)
   c. Weapon assembly

However, in a uranium-based program the chemical processing involves conversion to a single, specific uranium compound: uranium-hexaflouride (UF_6). Why this particular compound? As it turns out, all currently viable techniques employed for uranium isotope separation require that the input material be in a gaseous state—and UF_6 is the most practical choice.

In any case, the UF_6 (consisting of 99 percent $U\text{-}238F_6$ and 0.7 percent $U\text{-}235F_6$) is fed through the enriching facility and emerges as a final product of 95 percent $U\text{-}235F_6$, and a waste product of 99.8 percent $U\text{-}238F_6$.

The final product of 95 percent $U\text{-}235F_6$ is then reconverted, by chemical processing, into uranium metal. This metal, in direct contrast to the original natural uranium metal, is almost pure U-235.

The weapon fabrication team then fashions two subcritical masses of uranium metal (now composed of 95 percent U-235). The nonnuclear components of the weapon are then assembled, including the ballistic "gun" barrel, the chemical explosives, the fusing, detonators, wiring, and so forth. The U-235 masses are emplaced, and the device is sealed.

Having described how the program components fit together, let us now consider them individually and explain their corresponding indicators.

Mining, Milling, and U-Metal Conversion

The mining and milling stages provide the basic uranium compounds used in making the fuel for the production reactor. The annual ore demand would be roughly one hundred tonnes of U_3O_8 (equivalent), which in terms of actual ore tonnage could be anywhere from 2,500 tonnes (high-grade uraninite) to 500,000 tonnes (low-grade phosphate ores). Since the uranium content of these ores varies considerably within as well as between nations, the parameterization of the mining and milling stages was confined to the single value: one hundred tonnes of U_3O_8 equivalent.

The mining of uranium ores is not distinguishable from the mining of other mineral-bearing ores. In fact, uranium ores are almost always mined along with other minerals so as to increase the net economic return of the mining (Stephenson 1954, 323; Sanders 1975, 37). Thus the only characteristic resource demand components directly associated with the mining stages are the existence of uranium mineral deposits, some prior level of national activity in the mineral industry (Stephenson 1954, 323), and the associated capital costs of the uranium activity.

The milling of uranium ores also involves a standardized process in the mineral industry. The recipe for extracting and concentrating the U_3O_8 can be found in the open literature, as can data on design, construction, and operation of the milling facility. It has been estimated that a milling plant with an output capacity of about fifty-five tonnes of U_3O_8 per year could be designed and set up in less than two years by "any reasonably well trained metallurgist" (Lamarsh 1976, 10-11). For the posited milling plant, of approximately double capacity, the design and set-up demands would not be greatly different, but one would need an experienced chemical engineer. Finally, the required tools and machinery are available on the international market without restriction (Lamarsh 1976, 10-11). Thus the characteristic resource demand components associated with the milling stage include: an experienced metallurgist, an experienced chemical engineer, the primary construction materials—steel and concrete—and associated capital costs.

The U-metal conversion plant could very well be constructed concurrently with the milling plant and, for reasons of efficiency, would be located in the same place. The conversion of U_3O_8 into U-metal ingots is also based on standardized metallurgical processes found in the minerals industry.[2] Under the guidance of an experienced metallurgist and a chemist/chemical engineer, a U-metal conversion plant could be erected in less than two years. As with the milling plant, the necessary tools and machinery could be imported. The associated resource demand components for the U-metal plant therefore consist of: a metallurgist, a chemist/chemical engineer, construction materials, and capital costs. Of course the actual construction of the mill and U-metal conversion plant will require a small work force, perhaps as few as two hundred workers over a two-year period.

Fuel Fabrication

The fabrication of the fuel assemblies for the production reactor involves shaping the raw U-metal into "slugs," metal rods roughly one inch in diameter and several inches long. These slugs are then sealed inside hollow aluminum cartridges.[3] It has been estimated that it would take our already overworked metallurgist (or a technical assistant) about a year to fabricate the fuel assemblies.[4]

The posited fuel fabrication plant would be, in reality, little more than a common metalworking shop. Therefore, from a resource demand perspective: a metallurgist could easily design the facility; the construction materials are of little significance; and the associated capital costs would be very small.

The Production Reactor

The production reactor design might best be based on the Brookhaven Graphite Research Reactor. Detailed descriptions of its design and construction can be found in the open literature, though the design of the reactor could incorporate many shortcuts (Chastain 1958; U. S. Atomic Energy Commission 1955). A breakdown of the Brookhaven design is given in table 22.

Having already discussed the fuel-related research demands, I move on to the reactor's graphite requirements. Lamarsh (1976, 13) notes that reactor-grade graphite could conceivably be purchased on the international market. But, such a transaction—involving some seven hundred tonnes of nuclear graphite—would certainly attract the attention of other nations. The prospective nth country would be

TABLE 22
GENERAL DESCRIPTION OF THE BROOKHAVEN GRAPHITE RESEARCH
REACTOR

Power: 30 MWth[a]
Neutron Flux: 5 x 10_{12} (average thermal)[a]
Fuel: 84 tonnes of U metal in aluminum cartridges in a 33 ft x 33 ft loading[b]
Moderator: 670 tonnes of graphite[a]
Coolant: air at 300,000 cubic ft/min, with five blowers at 1,500 horsepower each
    requiring a total of 6,300 kw of electricity[c]
Shielding: (outward from core):
a. graphite reflector, 4.5 ft[d]
b. 6 inches of steel plate[e]
c. 4.5 ft of high-density concrete
d. 3 inches of steel for structural support
Control: steel alloy rods with a small fraction of boron[f]

---

[a]U. S. Atomic Energy Commission (1955, 3-4).
[b]U. S. Atomic Energy Commission (1955, 395).
[c]Chastain (1958, 156-57); U. S. Atomic Energy Commission (1955, 389).
[d]Lamarsh (1976, 24).
[e]U. S. Atomic Energy Commission (1955, 400-404).
[f]U. S. Atomic Energy Commission (1955, 388).

hard put to explain such a purchase. So for reasons of secrecy and security (i.e., possible preemption) it seems doubtful that an aspiring nuclear weapons country would prematurely broadcast its intentions in such a fashion. Moreover, as of 1974, reactor-grade graphite was incorporated into an IAEA "trigger list" for monitoring and safeguards (Sanders 1975, 62). Thus Lamarsh's assumption regarding market access to nuclear graphite has been weakened even further.

Disregarding the possibility of a "black market" purchase for now, it appears that a prospective nth country would have to produce its own graphite. In this context Lamarsh notes: "It should be mentioned that AGOT [reactor grade] graphite is similar to electrode graphite. Facilities for the manufacture of electrode graphite can easily be converted to production of reactor-grade graphite" (Lamarsh 1976; see also Currie, Hamister, and MacPherson 1955 and Nightengale 1962).

As shown in table 22, the cooling system would employ five 1,500 horsepower blowers, consuming some 6,300 kw of electricity, to move 300,000 cubic feet of air through the reactor every minute. The blowers and suitable air-duct systems are basic industrial equipment available on the international market (Lamarsh 1976, 13-14).

Finally, the shielding and control requirements—steel and concrete—are easily satisfied by indigenous capacity or market purchase.

The first reactor-related set of resource demand components represents the scientific and technological demands. Table 23 outlines a skeletal possibility. These manpower requirements are for supervisory personnel during the design and construction phases. For the actual operation of the production reactor, all that would be required is several of the scientists/engineers from the design team plus a staff of several tens of technicians (Beck 1957, 83). As for the educational requirements of the associated technicians: "it is generally agreed that for reactors operating on a round the clock schedule technicians with *no* formal engineering background serve most

TABLE 23:
SCIENTIFIC/ENGINEERING MANPOWER FOR REACTOR AND FUEL
SUPPORT ACTIVITIES

Profession	Number	Utilization
Civil structural	1	Structures, reactor building
Electrical	1	Control, circuitry, instrumentation
Mechanical	2	Heat transfer, mechanical devices
Nuclear	2	Design theory, etc.
Metallurgical	1	Uranium production
Chemical	1	Uranium production

Note: Based on calculations of Lamarsh (1976, 17-18).

efficiently" (Chastain 1958, 303; emphasis added). It seems, then, that the operating staff for the reactor would present no particular resource demand problem.

Likewise, a construction crew of one hundred workers has been estimated as adequate over a three- to four-year construction period (Lamarsh 1976, 17). Construction materials would be mainly concrete and steel. And of course a general "industrial engineering" team, encompassing the civil structural, electrical, and mechanical engineering specialties would be necessary. There would also be a demand for nuclear physicists, nuclear chemists, and nuclear engineers. The metallurgist and chemist "demands" have already been noted.

Finally, the associated capital costs of the production reactor itself should be noted.

Plutonium Reprocessing Plant

Perhaps one of the most contentious issues related to the feasibility of a plutonium-based weapons program for "small powers" is the plutonium reprocessing facility. Some claim such facilities are relatively cheap and simple to construct; others portray them as very sophisticated hurdles for prospective nth countries.

First, the construction and operation of a reprocessing plant is more a chemical engineering project than a nuclear engineering project (Groves 1962, 43). Second, *all* data regarding the design, construction, and operation of such facilities were declassified and distributed through the IAEA membership in 1955 (Goldschmidt 1977, 72). And third, the chemical and chemical-engineering principles involved in the design and operation of reprocessing plants are similar to, though more complex than, standard industrial practice (Stephenson 1954, 324-35; Glasstone 1955, 485-508.)

The complications arise because the material being processed is radioactive and therefore the plant operators must be shielded from the processing operations. However, the extremely low burn-up of the feed material (120 MWd/te) posited in the base case program greatly reduces this radiological hazard and correspondingly permits a number of design simplifications for the plant. In this context it is worth noting that the confusion regarding the difficulty of building reprocessing plants has arisen because some authors have failed to distinguish between commercial-scale reprocessing plants (those for handling large quantities of high burn-up power reactor fuel) and alternative low burn-up, low capacity reprocessing plants.[5] While spent power reactor fuel (7,000-33,000 MWd/te burn-up exposure) may exhibit radioactivity on the order of 2 million to 3 million curies per tonne, the low burn-up fuel would not have shielding requirements beyond those found in a typical research laboratory "hot cell."[6]

As for the reprocessing operation itself, the spent fuel could be reduced in simple batch operations conducted in concrete canyons.[7] The fuel rods could first be dissolved in an aqueous solution of sodium hydroxide and sodium nitrate, thereby removing the aluminum cladding (Bebbington 1976, 33). Then, employing the PUREX process (alternate baths of nitric acid and n-tributyl phosphate), the plutonium could be separated from the uranium (Glasstone 1955; 497-99; Jackson and Sadowski 1955). Further purification and conversion to pure plutonium metal could then be completed. Done in this fashion, a 95 percent recovery of plutonium could be expected (Bebbington 1976, 32).

In terms of quantity, the plant must be capable of "batch" processing only some eighty to ninety tonnes of very low burn-up reactor fuel in the course of a year (natural uranium metal at 120 MWd/te). Thus all the complications—the extensive plumbing, the massive shielding, the sophisticated electronics—encountered in a commercial-scale reprocessing plant are simply avoided (U. S. House of Representatives 1975, 96; Office of Technology Assessment 1977, 178). Primary construction materials include stainless steel and concrete, while the processing chemistry involves nitric acid and a variety of common organic compounds (Wick 1967; Groves 1962, 84-86; U. S. Atomic Energy Commission 1957).

The design and supervisory personnel for the reprocessing plant could be drawn from the same team that coordinated the construction of the other facilities: nuclear engineers, chemical engineers, metallurgists, and various industrial engineers. In all, fewer than a dozen such specialists would be required (Greenwood, Feiveson, and Taylor 1977, 146; Office of Technology Assessment 1977, 178). Approximately one hundred workers, employed over a three- to four-year period, could complete the facility. There are, to be sure, associated capital costs.

The Weapon Fabrication Laboratory

The final stage is, of course, the weapon fabrication stage. Here a nuclear physicist, working with a metallurgist, would design and fabricate the fissile core. Explosives and electronics experts, most likely military personnel, would assemble the nonnuclear components. In this regard it should be noted that the requisite electronics and (implosion) explosive techniques have long been incorporated into standard industrial process (Olgaard 1969, 219-20; United Nations 1968, 58-59; Wentz 1968, 20), and certainly into the realm of conventional military technology. Thus a prospective nth country could draw from the ranks of its military for the latter specialties. In sum, the weapons laboratory would require: nuclear specialists, metallurgists, explosives and electronics experts, and construction materials and capital for the facility itself.

Table 24 summarizes the various resource demand components associated with each of the program stages. Clearly, many of the resource demand components are redundant. In table 25 I have eliminated the redundancies and present the resource demand components as program components.

As is true of most data-based research, some empirical referents elude direct measurement. Unfortunately, this was true for some of the resource demand components listed in table 25. In particular, data for many of the manpower specialties listed there (e.g., nuclear engineers) are not uniformly available over the spatial and temporal

TABLE 24
RESOURCE DEMAND COMPONENTS BY STAGE

Stage	Resource Demand Components
1. Mining	Indigenous uranium deposits Previous national mining activity Initial operating costs
2. Milling	Metallurgists Chemical engineers Concrete, steel Construction force capital Research, development, and testing (RD&T) costs Initial operating costs
3. U-metal conversion	Metallurgists Chemical engineers Concrete, steel, electricity Construction force Capital RD&T costs Initial operating costs
4. Fuel fabrication plant	Metallurgist Electricity Capital RD&T costs Initial operating costs
5. Production reactor	Industrial engineers Nuclear engineers/physicists Metallurgists Chemical engineers Concrete, steel, electricity Graphite production capacity Construction force Capital RD&T costs Initial operating costs
6. Plutonium reprocessing plant	Chemical engineers Nuclear engineers Industrial engineers Metallurgists Concrete, steel, electricity Nitric acid Construction force Capital RD&T costs Initial operating costs
7. Weapons Fabrication Laboratory	Nuclear physicists Metallurgists Explosives/electronics experts Electricity Construction force Capital RD&T costs of weapon Initial operating costs

TABLE 25
PRELIMINARY LIST OF RESOURCE DEMAND COMPONENTS FOR THE BASE
CASE ATOMIC WEAPONS PROGRAM

Previous national mining activity
Indigenous uranium deposits
Metallurgists
Steel
Construction work force
Cement/concrete
Chemical engineers
Nitric acid (production capacity)
Electricity (production capacity)
Nuclear engineers/physicists/chemists
Nuclear graphite (production capacity)
Electronics/explosives specialists
Capital costs of various plant facilities
Research, development, testing, and engineering costs
Initial operating costs of the process plants
Industrial engineers: civil structural, electrical, mechanical specialties

domains of this study. For example, the science and technology manpower listing available from UNESCO since 1970 aggregates all engineering specialties into a single category. Similarly, physicists, geologists, astronomers, chemists, and so forth are lumped in another aggregate group. Consequently, surrogates were employed as indicators for some of the resource demand components.

A second issue is measurement. There are some resource demand components—for example, capital costs—that can be dealt with metrically. However, for a nuclear weapons program of the kind posited here, there are many resource demand components for which metric scaling conveys little useful information. These are resource demands for which possession of the qualitative capacity (either a country has it or it doesn't) implies satisfaction of the relevant quantitative demands. For instance, consider the resource demand component that reflects the capacity to produce nitric acid. We can safely assume that if a country produces nitric acid, then it produces enough of it to satisfy the program's quantitative demands. Thus, for those resource demand components for which qualitative satisfaction implied quantitative satisfaction, I used a supply threshold criterion.

Below, the indicators for each of the resource demand components listed in table 25 are discussed in detail. Some are self-explanatory and therefore need no evaluation. Others, however, are not as intuitive and so require some development.

Previous National Mining Activity

Indicator (d_1):   some fraction of the country's labor force was engaged in mining activities.

Data source: *United Nations Statistical Yearbook.*

Indigenous Uranium Deposits

Indicator (d_2):   Known uranium deposits within the national territory.

Data source: *United Nations Statistical Yearbook*, Barnaby (1969a), U.S. Senate (1975, 69-100), Stockholm International Peace Research Institute 1974.

Metallurgists

A surrogate indicator is required for this resource demand component. Uranium metal possesses metallurgical properties comparable to those of several common metals. These include austenitic stainless steel, copper, and magnesium (Wilkinson 1962). When personnel were drawn from existing pools of "conventionally" schooled metallurgists, training metallurgists for the nuclear industries of Britain and France (at levels of nuclear development comparable to that facing a prospective nth country today) involved three to four months of supplementary education (Organization for European Economic Cooperation 1956, 69-70).

Accordingly, we can assume that substantial experience with indigenous steel production implies that at least a small pool of metallurgists is likely to be available for a government-sponsored nuclear weapons program.

Indicator (d_3):   since attaining national independence, the country had been producing steel in indigenous facilities for at least Y years.[8]

Data source: *United Nations Statistical Yearbook.*

Since indigenous steel production was used as a surrogate for the metallurgist component, the need for an indicator of steel production capacity (as listed in table 25) has been met. (Of course the steel itself could also have been purchased on the international market.)

Construction Work Force

A surrogate indicator is also needed for this resource demand component. The total work force required for the base case nuclear weapons program would be under one thousand. Thus, in terms of raw manpower this component seems trivial. However, the construction work force would have to be familiar with construction techniques with modern materials—especially steel and concrete. In this sense an indigenous production capacity for both steel and cement suggested a high likelihood that there was some degree of national activity in industrial construction. Since a steel production capacity was already noted, I recorded cement production capacity.

> Indicator (d_4):   since attaining national independence, the
> country had been producing cement in indig-
> enous facilities for at least Y years.
>
> Data source: *United Nations Statistical Yearbook.*

Since indigenous cement production was used as a partial surrogate for the construction work force component, the need for an indicator of cement production capacity has been met. (Of course the cement itself could also have been purchased on the international market.)

Chemical Engineers

Here too a surrogate is used. A good indicator of national chemical-related industrial activity is the production of acids, which in turn implies national access to chemical engineering talent.

Since two of the most frequently used acids are nitric acid and sulfuric acid, I assumed that the indigenous production of *either* acid (over a period of time) implied a high relative likelihood that at least a small pool of chemical engineers would be available for a government-sponsored nuclear weapons program.

> Indicator (d_5):   since attaining national independence, the
> country had produced nitric acid for at least
> Y years.

Or

> Indicator (d_6):   since attaining national independence, the
> country had produced sulfuric acid for at
> least Y years.
>
> Data source: *United Nations Statistical Yearbook.*

Nitric Acid (Production Capacity)

Of course the surrogate indicator d_5 is also the appropriate indicator for this resource demand component. But a nation that had experience in producing other acids and also had facilities for producing (nonorganic) nitrogenous fertilizers could quickly begin to produce nitric acid. (In fact, nonorganic nitrogenous fertilizers and nitric acid are often produced in tandem.) Therefore, even if a country was not currently producing nitric acid, it was assumed to be capable of doing so if it had reasonable experience in sulfuric acid production (d_6) *and* nonorganic nitrogenous fertilizer production.

Indicator (d_7):   since attaining national independence, the country had produced nonorganic nitrogenous fertilizers for Y years.

Data source: *United Nations Statistical Yearbook.*

Electrical Production Capacity

The total electrical power demands of the base case program posited in this study were determined to be under 10,000 kilowatts (kw). (By comparison, it has been estimated that the Hanford plants consumed over 100,000 kw.) Assuming that no more than 5 percent of the nation's electric power production capacity would be available for "diversion" to an atomic weapons program, this suggested that a prospective nth country had to possess an installed capacity of at least 200 megawatts electric (MWe).[9] Moreover, in order to "tinker" with the power grid, some minimal working experience at that power level was presumed necessary.

Indicator (d_8):   since attaining national independence, the country had installed an electrical capacity of at least 200 MWe (about 1.8 billion kilowatt-hours/year) for at least Y years.

Data source: *United Nations Statistical Yearbook.*

Nuclear Engineers, Physicists, Chemists

A surrogate indicator was required for this resource demand component. The indicator had to reflect both manpower availability and acquaintance with the necessary scientific and technological information. As a report by the United States Office of Technology Assessment observed (1977, 178), this group of specialists must have had "applicable practical experience. A reactor and reprocessing plant cannot be built by reading books alone." The operation of a research reactor (and its associated facilities, e.g., a "hot cell") connotes such experience. Experiments designed to determine the various parameters associated with assembling a critical mass so as to produce a fissile explosive yield could be casually conducted. Given the diverse number of experiments that would probably be required, approximately three research-reactor-years of operating experience would be needed before the final assembly of the weapon components.

Research-reactor-years of experience was computed as:

$$E(t) = \sum_{i=1}^{n} e_i(t),$$

$E(t)$ = research-reactor-years of experience as of the year t,

$e_i(t)$ = number of years that the ith research reactor had been in operation as of the year t, for a country with n research reactors.

Indicator (d_9): since attaining national independence, the country had at least three research-reactor-years by the year its nuclear weapons program schedule would have it assemble the first nuclear weapon. This indicator can be computed as: year (d_9) = $T + 3 - N$; where T is the year in which three research-reactor-years are attained and N is the nominal length of the initial phase of the nuclear weapons program.

Data source: *United Nations Statistical Yearbook.*

Nuclear Graphite (Production Capacity)

A surrogate indicator is required for this resource demand component. "Most electrographites are produced from a petroleum-coke filler and a coal-tar pitch binder. Nuclear grades of electrographites are produced by modifications of conventional manufacturing methods, with special care being taken to exclude impurities that have significant neutron-absorption cross-sections" (Eatherly and Piper 1962, 21; see also Currie, Hamister, and MacPherson 1955; Nightingale 1962). Thus the basic capability to produce nuclear graphite can be indicated by petroleum coke production and coal processing. On the one hand, petroleum coke is a normal by-product of petroleum refining. However, many nations that could easily produce petroleum coke simply do not because the demand for the substance is relatively low. Nonetheless, petroleum coke production could be quickly initiated by the thermal "cracking" of heavy oil distillates at a petroleum refinery (Kirk and Othner 1949, 3:4-5, 10:192-202). On the other hand, the production of coal tar pitch is an intermediate step in coal coking (Kirk and Othner 1949, 3:4-5). Thus I assumed that a nation that engaged in petroleum distillation (the production of kerosene, naphtha, or heavy oils) and coal coking could make the necessary modifications for the production of nuclear graphite.[10]

Indicator (d_{10}): since attaining national independence, the country had been distilling petroleum for at least Y years.

And

Indicator (d_{11}): since attaining national independence, the country had been engaged in coal coking for at least Y years.

Data source: *United Nations Statistical Yearbook.*

Electronics/Explosives Specialists

A surrogate indicator is used for the *electronics* specialist resource demand component. The electronics specialist would have to be acquainted with the methods and techniques employed in electronic ignition devices—in order to cope with the relatively unsophisticated design/construction of the fusing/detonator system. There is in many ways an interesting correspondence between this system and the electrical ignition systems used in internal combustion engines (though not a one-to-one correspondence). To tap this need for experience with electromechanical ignition systems, I assumed that experience with motor vehicle manufacture or with automotive assembly plus radio (or television) production implied a high likelihood that a fusing/detonator system could be constructed.

Indicator (d_{12}): since attaining national independence, the country had been manufacturing motor vehicles for at least Y years.

Or

Indicator (d_{13}): since attaining national independence, the country had been assembling motor vehicles for at least Y years.

And

Indicator (d_{14}): since attaining national independence, the country had been manufacturing radios (or televisions) for at least Y years.

Data source: *United Nations Statistical Yearbook.*

Reiterating an earlier point, any nation fitting the industrial profile above would certainly possess a small but sufficient pool of explosives/demolitions specialists within its armed forces or its industrial work force (United Nations 1968, 58-59; Wentz 1968, 20; Meyer 1978d, 238).

Setting the Value of Y

What remains is to derive a value for Y, the period of time used to denote "experience" with industrial processes. The objective is to use time as a surrogate for the "Sears, Roebuck" factor in the analysis. This involves an iterative strategy. Initially, I set Y at ten years—a nice round number. Then, using the fourteen indicators (with Y equal to 10.0), I calculated the earliest date that each nation satisfied the various resource demands and correlated this with the "dates" found in historical studies of various nations. With Y set at 10.0, the indicators tended to overestimate the historical dates by several years. Next I set Y at 5.0 and made new comparisons with the historical data. With Y set at 5.0, the indicators tended to underestimate the historical dates. Using Y equal to 8.0 (an approximate mean value), I found substantial agreement between the indicators and the historical data.[11] (Curiously, a value of 8.0 has a substantive connotation. It reflects roughly two generations of college graduates.)

The scientific, technical, and industrial components, along with their respective indicators, are summarized in table 26. They in

TABLE 26
SUMMARY OF RESOURCE DEMAND COMPONENTS AND THEIR INDICATORS

Resource Demand Component	Indicator
Previous national mining activity	d_1
Indigenous uranium deposits	d_2
Metallurgists	d_3
Steel	d_3
Construction work force	$d_3 + d_4$
Cement	d_4
Chemical engineers	d_5 or d_6
Nitric acid	d_5, or $d_6 + d_7$
Electricity (production capacity)	d_8
Nuclear engineers, physicists, chemists	d_9
Nuclear graphite (production capacity)	$d_{10} + d_{11}$
Electronics specialists	d_{12}, or $d_{13} + d_{14}$
Explosives specialists	Σd_r

effect represent the first "dimension" of the control variable. Again, in selecting the various indicators, I made an effort to capture the characteristics of each nation's underlying scientific, technological, and industrial infrastructure as well as the resource components specifically linked to the manufacture of atomic weapons. Thus the obvious need for carpenters, plumbers, masons, electricians, and so on can be assumed to be implicitly satisfied if all the particular resource demands are themselves satisfied.

If a nation can satisfy all the scientific, technological, and industrial resource demands implied by the indicators d_1 through d_{14}, then it should be capable of undertaking any one of the *individual* activities that constitute an atomic weapons program. However, acquiring the capability to manufacture atomic weapons is predicated on two conditions: (1) being able to undertake every one of the requisite program activities, and (2) being able to pursue those activities in a coherent and coordinated manner—that is, as a program. In this respect, being able to undertake any or all of those activities as singular products implies nothing about being able to pursue them within an integrated program. *Thus, satisfying all the scientific, technological, and industrial resource demands described here is a necessary component of a latent capacity to manufacture nuclear weapons.* The other necessary component—the basic ability to pursue the various activities as an integrated program—is derived in Appendix C in terms of economic resource demand components.

Appendix C
Cost-Estimating Procedures

Even the casual reader of the nuclear proliferation literature will be aware that there is a wild divergence in cost estimates for an atomic weapons program. In part this is due to differing assumptions regarding program size and scope. Some researchers can think only in terms of large-scale, high-technology approaches involving massive gaseous diffusion plants or giant production reactors serviced by fully automated reprocessing plants. Others prefer to dwell on "quick and dirty" approaches—one-shot processes capable of producing only a single weapon. Adding to the confusion are the confounding influences of civil nuclear technologies. All too many scholars have incorrectly equated the process facilities needed for a nuclear weapons program with those used in commercial (civil nuclear) power programs. This has been particularly true with respect to nuclear reactors and reprocessing plants. The fallacy here cannot be overemphasized. The design and operations complexities of plants for an atomic weapons program, compared with those for civil nuclear programs, are an order of magnitude (or more) smaller.

To be sure, the capital, or investment, costs for the various plant facilities tend to vary with location and process type. However, the cost uncertainties for the small facilities posited here are substantially less than those associated with more elaborate commercial-scale plants.

All the cost estimates used in this study are based on the technology of the 1950s to mid-1960s. I have purposely chosen "old technologies" because they are reliable (low risk with respect to successful operation), well described in the literature, and relatively inexpensive, and because they seem to represent a logical starting point for a prospective nth country to build upon.

I made an explicit effort to gather data that reflected a wide variety of national and engineering experience. Thus these cost estimates are mean values. As such they may tend to overstate actual costs—since prospective nth countries could opt to build "bare-bones" facilities (see especially my discussion of plutonium reprocessing plants). Nonetheless, these estimates do provide a good basis for assessing the capital costs most likely (in a probabilistic sense) to be incurred in pursuing an atomic weapons program of the kind outlined in chapter 2.

Finally, let me draw attention to the size of the scale factors used. By traditional industrial experience they may indeed seem small. However, keep in mind that our posited plants are not industrial size—they are pilot-scale plants. Moreover, for all the chemical processing facilities (including the reprocessing plant), the investment costs are determined not so much by the process equipment as by the absolute size of the plants themselves. Thus the cost variance among pilot-scale plants should be expected to be small, since "economies of scale" do not become manifest at the pilot-scale level (i.e., initial "buy-in" costs are fixed).

Capital Costs of Uranium (d_{15})

The capital costs associated with opening a uranium mine are extremely sensitive to the location of the ore deposits, the ore type, its assay value, and the relative location of other valuable minerals. Consequently, investment costs could vary from a few hundred thousand dollars to several million dollars.[1] Therefore, for the sake of its nominal inclusion I put the cost at 1.5 million.

Capital Costs of Milling (d_{16})

Parameters: *a.* average ore assay of 0.2 percent uranium
b. 100 te U_3O_8 per year plant output capacity

Assuming a 10 percent material loss during processing, the base case program would require a mill with an annual ore input capacity of 55,000 tonnes. In this respect the Spanish mill at Andujar (capacity equal to 60,000 te/yr) reported costs of about $1.9 million (1960 dollars).[2] A substantially larger Canadian mill with a 440,000 te/yr capacity cost some $9.0 million.[3] Applying, then, a 0.70 scaling rule as reported in a 1968 study for the United States Atomic Energy Commission, the 55,000 te/yr plant should cost between $1.8 million and $2.1 million (Little 1968, 169).

195

A mill capacity versus mill cost analysis was conducted explicitly for this study (Meyer 1978a). Data for some ten mills of varying size, location, and construction date were compared. Based on that regression analysis, a 55,000 te/yr uranium ore mill should cost about $1.6 million.

(Note that the exact cost of a uranium mill will depend on the type of ore being processed. The figures suggested here are for ores of average quality. For an interesting example, see Wordsworth 1969.)

Finally, the USAEC reports the median cost of a 55,000 te/yr mill as $1.9 million (1966 dollars), or $1.7 million in 1960 dollars.

Using an averaged cost, the 55,000 te/yr mill could run about $2 million (1960 dollars).

Capital Cost of the Conversion and U-Metal Plants (d_{17})

      Parameters: *a*. 85 te U-metal per year output
                       b. 300 days per year (equivalent) operation

The U_3O_8-concentrate to U-metal conversion process consists of several distinct phases:
    1. $U_3O_8 \rightarrow UO_3$
    2. $UO_3 \rightarrow UO_2$
    3. $UO_2 \rightarrow$ U-metal

Using the FUELCO model developed by the USAEC, a processing facility for the first phase should cost about $1.5 million (1965 dollars). The same model estimtes $0.7 million for the second phase, and $0.9 million for the third. Thus the total cost would be $3.0 million (1960 dollars).

An independent Organization for European Economic Cooperation estimate for a similar grouping of uranium processing plants, based on a 450 te/yr capacity and scaled down to a 85 te/yr capacity, produced a figure of $2.7 million (1960 dollars).[4]

Finally, as a consistency check, I compared the estimate for the base case program requiring 85 te/yr of U-metal with that for an analogous program posited by a United Nations committee (United Nations 1968). Their program, using a heavy-water moderated reactor, called for a 20 te/yr uranium processing capacity with a total cost of $2.5 million (1967 dollars) and including a uranium mill. If a small mill would cost $0.6 million (1967 dollars), then conversion and U-metal plants would cost about $1.9 million. Using a weighted average scale factor for the U_3O_8 U-metal facilities, I calculate that the 20 te/yr plants would total $1.9 million (1967 dollars).[5] Thus it seems that my cost estimate for the 85 te/yr conversion and U-metal plants is consistent with those of other studies with different specifications.

Reiterating, I have adopted a figure of $3 million (1960 dollars) for the capital costs related to the U_3O_8 U-metal conversion process.

Capital Costs of the Fuel Fabrication Plant (d_{18})

Parameters: *a.* 85 te/yr capacity
b. U-metal fuel clad in aluminum

A simple yet functional fuel fabrication plant for canning U-metal in aluminum is best typified by the Indian plant built in 1959. Costing some $1.0 million, it has a nominal capacity of 30 te/yr (Prakesh and Rao 1961, 392-402). A somewhat more elaborate facility, capable of handling about 200 te/yr, has been estimated to cost $2.9 million (1960 dollars).[6] Based on studies conducted by researchers at Oak Ridge, natural uranium fuel fabrication plants scale in accordance with a 0.3 exponent (Culler 1963, 14). Thus, our 85 te/yr fuel fabrication plant could cost between $1.4 million and $2.2 million.

Remember that the type of fuel being fabricated in the posited atomic weapons program is of very elementary design and is not comparable to that used in modern nuclear power reactors. Also, apparent cost discrepancies sometimes arise because some fuel fabrication plants incorporate some of the chemical processing stages that I have considered separately.[7]

In this context, a French estimate for a simple "canning facility"—which most closely resembles the one posited in our program—came to $1.2 million (1960 dollars) (Andriot and Gaussens 1958). Again the variance is due to facility versatility and engineering considerations. The average cost, and the one I used, was about $2 million (1960 dollars).

Capital Costs of the Production Reactor (d_{19})

Parameters: *a.* 30 MWth power rating
b. air-cooled, graphite moderated
c. natural U-metal fuel in aluminum cladding

An excellent cost analysis of a "stripped down" version of the Brookhaven 30 MWth research reactor was done by Lamarsh (1976). He derived an approximate cost somewhere between $13 million and $26 million (1975 dollars). Averaged, and converted to 1960 dollars, such a reactor might be expected to cost $11 million.

The actual costs of the Brookhaven reactor are as follows (U. S. Atomic Energy Commission 1955, 389):

Reactor	$9.3 million
Building	3.0 million
Air cooling	3.8 million
Total	$16.1 million (1950 dollars)

This corresponds to $21 million in 1960 dollars.

Capital Costs of the Plutonium Reprocessing Plant (d_{20})

Parameters: *a*. 85 te/yr uranium input capacity
b. batch processing
c. PUREX process

Here it is extremely important to make the distinction between reprocessing plants designed for high burnup fuels and those designed for low burnup fuels. A facility to treat low burnup fuels would require less than half the capital investment needed for an equivalent high burnup facility (Guthrie 1957; U. S. Atomic Energy Commission 1957, 1971). Since the base case program involves fuel exposures of less than 150 MWd/te, let me emphasize that this discussion involves only low burnup reprocessing plants.

Since 1955 an extraordinary amount of data pertaining to the design, contruction, and cost of small-scale, low burnup, reprocessing plants has become available. This availability of information, coupled with the fact that a (low burnup) plutonium reprocessing plant is (with the exception of a moderate degree of radiation shielding) not unlike other types of chemical processing plants, suggests that the associated research and development costs would be minimal.

The Indian facility at Trombay, nominally designed for a 30 te/yr throughput, reportedly cost Rs 35 million, or $4.8 million (1960 dollars). Pushed to its operational limit, however, it could handle between 60 and 90 te/yr of natural uranium fuel (India, Department of Atomic Energy 1965, 21). The more versatile Hanford Semi-Hot Works facility, with a capacity of about 60 te/yr, required a capital investment of some $8.8 million (1960 dollars). However, the small Hanford plant was a "pioneer" facility at the time it was constructed (1944-45); today an identical plant would cost substantially less (in constant dollars).

Nonetheless, using this information, I was able to recalibrate the FUELCO model (which is set up for commercial processing facilities) to make it applicable to low burnup plants (U. S. Atomic Energy Commission 1971). Accordingly, it seems that an 85 te/yr plant would cost some $8.2 million (1960 dollars). This estimate is consistent with the scaling rules noted by a number of researchers.[8]

In actuality, a "stripped down" facility could probably be built for much less than $8.2 million. Consider the two-phase procedure for producing plutonium metal: (1) separation of the plutonium from the uranium fuel element; (2) fabrication of plutonium metal ingots. A very simple direct-maintenance facility for producing concentrated plutonium solutions could be modeled after the well-described ORNL metal recovery plant (Jackson and Sadowsky 1955; Lewis 1955). Designed to handle fuel from the ORNL X-10 reactor (which is identical to the posited production reactor except that it has about one-third the thermal power), the metal recovery plant can process over 100 te/yr of low burnup fuel. Its cost: $850,000 (1950 dollars)— $1.1 million in 1960 dollars.

The end products are an aqueous plutonium solution, an aqueous uranium solution, and a waste solution.[9] A second facility, a plutonium metal plant, would then convert the aqueous plutonium into plutonium metal—by the so-called bomb reduction method (Baker 1946). Owing to its small size and simple chemistry, this plutonium metal plant could cost less than $500,000 to construct.[10] Auxiliary facilities like waste-processing and storage, ventilation, and so on, might add some $1.1 million (Bosch, Oszusky, and Asyee 1968). Unlike commercial reprocessing facilities, no effort would be made to recover the uranium metal; thus keeping costs to a minimum. Thus, taken all together, this stripped down plant might cost some $2.7 million.

To be sure, we end up with two very disparate investment cost estimates. Again, the discrepancy is the result of different engineering concepts. Accordingly, while the lower estimate ($2.7 million) represented a "rock bottom" possibility, it seemed that the higher cost estimate ($8.2 million) reflected a more likely investment target. Reflecting a conservative bias, then, I adopted the figure of $8 million (1960 dollars) as the cost of our reprocessing plant, an averaged cost based on the Indian experience and the FUELCO estimate (appropriately calibrated).[11]

RDT&E Costs of the Process Plants (d_{21})

It is interesting to note that prior studies have steered clear of estimating Research, Development, Testing, and Engineering (RDT&E) costs. At first glance this might seem a wise decision, given the relative lack of information concerning this aspect of atomic weapons production. After all, an estimate of uncertain accuracy would only "contaminate" the otherwise "clean" figures. However, this argument ignores the fact that by totally excluding RDT&E costs, even greater bias may be imparted to the overall estimate than if some allowance (however inaccurate) was made for their inclusion. Accordingly, I have attempted to estimate the relevant RDT&E costs with what little information is available.

Several points should be noted. First, since the mid-1950s an ever-growing quantity of research reports, conference papers, engineering monographs, and such have become available on an unrestricted basis. Second, international conferences, exchanges, and training programs—methodically covering the many facets of nuclear science and nuclear engineering—have most certainly had a major impact on the extent of the RDT&E effort required toward the manufacture of atomic weapons.[12] It seems, then, that in general the combined effect of these considerations would be to make the RDT&E effort more time consuming than capital consuming.

Moreover, the small size of the base case program used in this study in many ways reduces the need for extensive RDT&E with respect to the process facilities themselves (Van Cleave 1974, 56). Indeed, they could conceivably be viewed as RDT&E projects in and of themselves. For this resource demand (cost) component I used a percentage formula, as opposed to an actual value, because the information available on these particular costs was extremely limited.

Cost breakdowns of the British program revealed that the RDT&E costs were of the order of 28 percent (actually between 25 percent and 30 percent) of the investment costs for the various plant facilities (Gowing 1974b, chaps. 14, 18). A comparable comparison of the individual cost components of the French program placed these RDT&E costs somewhere between 14 percent and 19 percent (average 16 percent) of the capital outlay for fissile material production.[13]

Here I should note that the French were able to benefit from the release of considerable amounts of nuclear technology data after 1953. That is to say, the French nuclear weapons program occurred farther down the nuclear learning curve. Therefore we might expect a reduction in R&D costs.

Based on an analysis of British and French RDT&E expenditures, and assuming a sine function learning curve, I estimated that the R&D costs would approximate:

$$0.12 \sum_r d_r;$$

for r = 16, 17, 18, 19, 20.

(Explicit consideration of learning curve effects is discussed below.)

Keep in mind that, while the inclusion of this cost component will reduce the overall bias of the total cost estimate (relative to the estimates of others who do not include RDT&E costs), as an individual component it may tend to overestimate/underestimate the "true value" of RDT&E costs.

Initial Operating Costs of the Process Plants (d_{22})

Quite clearly, certain costs will be incurred in attaining an initial operating capability; specifically, these are the costs of processing materials for the plants (e.g., the various solvents) and the costs of actually running the plants (e.g., electricity). For a program of the size considered in this study, such costs could equal as much as 20 percent to 25 percent of the capital investment.[14] Thus, operating costs were estimated by:

$$0.23 \sum_{r}^{1} d_r;$$

for r = 16, 17, 18, 19, 20.

Cost of Weapons Fabrication (d_{23})

This last resource demand component subsumes those costs that are associated with the design and manufacture of the atomic weapons themselves. These include the capital cost of the weapons "laboratory," the RDT&E costs incurred in the design phases of the program, and the cost of the nonnuclear components.

There is, to be sure, little concrete guidance. An Indian study reports that the total cost for the design and manufacture of a single plutonium weapon would be about $6.4 million (1975 dollars), or roughly 25% of their estimated expenditure for the plutonium core (Seshagiri 1975, 69). Correspondingly, there is also some evidence that the British and French expenditures for weapons design and manufacture fell somewhere between 20 percent and 25 percent of their respective expenditures for plutonium production (Gowing 1974, chaps. 14, 18; United Nations 1968, 62; Hohenemser 1962, 240; Dhar 1957, 148).

As the RDT&E costs, and the cost of the weapons laboratory, should be insensitive to differences in program size less than order of magnitude (i.e., weapons output), we would expect that the 20 percent to 25 percent of plutonium production cost fraction would be applicable to our posited program (United Nations 1968, annex IV), approximated by:

$$0.23 \sum_r d_r;$$

for $r = 16, 17, 18, 19, 20, 21, 22.$

Time-Dependent Cost Adjustments

Pegged to the year 1960, the total cost of the program summed to $61 million. However, for years other than 1960, specific time-dependent cost adjustments had to be made. Obviously there was a need to incorporate an adjustment for the effects of what might be termed the "international nuclear learning curve."

I posited that the cost of "going nuclear" before 1955 should have been greater than in the years directly after 1955 (in constant dollars). In a sense 1955 was a watershed year for the international diffusion of nuclear science and nuclear technology, in that the veil of secrecy that surrounded many aspects of existing national nuclear technology programs was lifted. In the light of the international "Atoms for Peace" movement, the wholesale exchange of nuclear-related information in science and technology occurred at both bilateral and multilateral levels. Reviewing the pace and scope of these international exchanges, the creation of a number of international organizations devoted to the exploitation of the peaceful atom, and the general rise of a host of national nuclear energy programs, it was evident that movement along the international nuclear learning curve was most pronounced between 1955 and 1960. The cost impact of this curve, in a similar fashion, should have been particularly manifest during this period.

Learning curves in general, and technological learning curves in particular, are characteristically portrayed as S-curves, in which the cumulative amount of new information absorbed is plotted as a function of time. The preliminary introduction of a new technology is marked by a period of slow initial learning. As the basic information is mastered, the rest of the material is assimilated more rapidly. The curve finally tails off as "fine points" are accumulated and stored.

Intuitively, we should expect that the cost impact of this learning curve at any specific point in time would be directly related to the total amount of information assimilated up to, and including, that time. In other words, each increment of new information contributes to a reduction in program costs (particularly with respect to RDT&E). But this effect is not constant over time; it is most pronounced during the middle periods.

As the result of a fortunate historical "coincidence"—the fundamental technological and production symmetries between the initial British and French atomic weapons programs—it was possible to use simple algebra to approximate the learning curve effect.

Before 1955:

$$2.4 \sum d_r(1960)$$

1955 to 1960:

$$2.4 \sum d_r(1960) \cdot \left(1 - .42 \sin 15 \left[\frac{\text{Year-1955}}{6} \right]\right)$$

After 1960:

$$\sum d_r(1960)$$

in constant 1960 dollars.

Again, these cost estimates are just that: estimates. They constitute a basis for making probabilistic assessments regarding the partial ability of prospective nth countries to meet the capital-type resource demands of a small atomic weapons program. Idiosyncratic cost variances across national boundaries owing to the availability of goods, the cost of labor, and so on, were *assumed* to randomly cancel out.

Appendix D
Estimating Nuclear Propensities

This appendix presents, in symbolic form, the procedures for estimating aggregate nuclear propensities as outlined in chapter 5.

The simple nuclear propensity associated with a single motive condition (in the absence of any dissuasive conditions) is given by:

$$p_i = \frac{\#\ (d \cap m_i) + 2}{\#\ m_i + 3} \tag{1}$$

where p_i is the simple nuclear propensity associated with the ith motive condition (m_i). The numerator is the number of instances where proliferation decisions (d) were systematically coincident with motive condition m_i, and the denominator is the total number of instances in which m_i is observed over the entire data set when no dissuasive conditions were present). Notice the inclusion of the Bayesian priors [2,3].

When more than one motive condition is present, but no dissuasive conditions, a simple product formula can be used. Given the independence assumption or the influences of motive conditions, we can estimate compound nuclear propensity using joint likelihood calculations. The compound nuclear propensity can be computed as:

$$P = 1 - \prod_{i}^{I} (1 - p_i),$$

where P is the compound nuclear propensity resulting from the simultaneous presence of I motive conditions, p_i is the simple nuclear propensity associated with the ith motive condition (m_i), and $\prod$ is a product operator.

The effects of dissuasive conditions on motive conditions are reflected in a revised nuclear propensity. This is given by:

$$P_{ij} = \frac{\#\ (d \cap m_i \cap n_j) + 1}{\#\ m_i \cap n_j + 3},\qquad (2)$$

where P_{ij} is the revised nuclear propensity for the ith motive condition (m_j) in conjunction with the jth dissuasive condition (n_j). The numerator is the number of instances in which proliferation decisions are coincident with m_i and n_j together, and the denominator is the joint occurrence rate of m_i and n_j across the data set. Here too Bayesian priors [1,3] are inserted into the equation.

In calculating the final aggregate nuclear propensity, one first computes the adjusted nuclear propensity:

$$p_{i*} = p_i \prod_j^J (p_{ij}/p_i),$$

yielding Pi^*, the adjusted nuclear propensity for the ith motive condition across J individual dissuasive conditions. Then, P^*, the aggregate nuclear propensity is computed by:

$$P^* = 1 - \prod_i^I (1 - p_i^*).$$

Notes

1. An explicit definition of latent capacity follows later in the discussion. For the present, latent capacity connotes sufficient technical, industrial, material, and financial resources to support a wholly indigenous nuclear weapons program.

2. For some of the clearer discussions see Dunn and Kahn (1976) and Wohlstetter et al. (1979).

3. For a discussion of Swedish research see Garris (1972). With respect to Pakistan see Dunn (1982, 44-48) and Potter (1982, 157-61). I recognize that there are other ways to "get the bomb"—purchase and theft being two possibilities. As I have explained, this preliminary analysis is restricted to indigenous production routes to acquiring the bomb; later chapters will broaden the scope of the analysis to include these other methods.

4. This study leans heavily on such case studies for basic data and theoretical guidance. See, for example, Kapur (1976), Gowing (1974), Scheinman (1965), and Kohl (1971).

5. An alternative interpretation of this is that the incentives for nuclear weaponry are almost universal. Therefore only technological opportunity determines whether nuclear proliferation occurs.

6. Shapely (1978, 155). Dr. York, formerly an atomic weapons scientist, is at present director of the Program on Science, Technology, and Public Affairs at the University of California at San Diego.

7. It should not be very suprising to discover that many proliferation forecasts did indeed take such a stance. See National
Planning Association (1960), Wentz (1968).

8. Remember that a proliferation decision may occur in the
absence of technological capability. But I am restricting the
discussion, temporarily, to cases where the technological
capability for indigenous nuclear weapons production is in
hand.

9. These may very well include bureaucratic/domestic (political)
costs and benefits that can become attached to the nuclear
issue. They may even be the primary incentives/disincen-
tives. Thus, this type of decision making process need not
necessarily conform with the well-known "rational-actor"
model.

CHAPTER 2

1. Among many useful works see Office of Technology Assessment (1977), Wohlstetter et al. (1978). Descriptions of nuclear
weapons design can be found in Seshagiri (1975, chap. 3),
McPhee (1974, 189-219) and Brown and McDonald (1977).

2. The historical record suggests that the United Kingdom made
two independent decisions to develop nuclear weapons. The
first British "program" was from 1939 to 1945 and was
eventually subsumed within the American Manhattan Project.
A second "program" was initiated in 1947 and culminated in
Britain's first atomic test in October 1952. The discussion
here refers to the second British atomic bomb program and is
drawn heavily from Gowing (1974), Pierre (1972), Rosecrance
(1964b, c), and Groom (1974).

3. The assumptions made regarding each of these three considerations were, to say the least, arbitrary. For a detailed
analysis see Gowing (1974), especially chapters 6 and 7.

4. In 1951 the government authorized the upgrading of the low-
enrichment facility to a high-enrichment facility.

5. It should be noted that the scope of the initial British program
was limited to plutonium-based weapons owing to the lower
associated resource demands.

6. See *New York Times* (8/9/77, 9), (8/2/77, 3), (8/23/77, 1), (8/29/
77), (8/31/77, 2).

7. "Moderate confidence," refers to the probability that a
nuclear explosion would occur. There might be considerable

uncertainty as to the precise yield. Also, the recent spate of senior honors projects by undergraduate physics students that are reputed to have produced workable atomic bomb designs provides some food for thought.

8. For an interesting discussion on the relation between warning and surprise, see Belden (1977). For a discussion of politico-military reasons *not* to test, see Wentz (1968, 34-41).

9. Note that the United States did not bother to develop a special delivery vehicle for its first atomic weapons.

10. Testimony by former nuclear weapons designer Theodore B. Taylor before the House Subcommittee on International Affairs. See United States House of Representatives (1975, 84).

11. Even the transcontinental United Kingdom/France versus Soviet nuclear rivalry may be an inappropriate model for analyzing future nuclear rivalries at the local level.

12. U.S. Arms Control and Disarmament Agency (1977, 155-80). The historical relationship between PNES and nonproliferation efforts can be deduced by examining the texts of the Nuclear Non-Proliferation Treaty and the Treaty for the Prohibition of Nuclear Weapons in Latin America.

13. The present state of controlled fusion energy production research is a good illustration of this dichotomy.

14. Again, we must recognize that many specific materials (e.g., stainless steel) and, indeed, entire plants could be purchased on the international market. For the sake of a more complete discussion, however, such alternatives can be ignored for now.

15. To be sure, there is an extensive body of literature that examines this issue from both theoretical and empirical perspectives. See, for example, Enthoven and Smith (1971), Hitch and McKean (1967), Russett (1970).

16. See, for example, Halperin (1974). Of course this says nothing about the eventual level of atomic weapons expenditures that a nation may tolerate. In particular, while the United States, the United Kingdom, and France began with initial allocations of 0.6, 1.2, and 1.2 percent of their respective annual military expenditures for atomic weaponry, in the ensuing years atomic weapons expenditures grew to consume between 10 percent and 20 percent of their defense budgets. So the distinction between the initial phase of the nuclear weapons program and the later phases of development must indeed be emphasized.

CHAPTER 3

1. For some interesting historical case studies see Gowing (1974, 407), Joshua and Hahn (1973, 15), Kohl (1971, 38, 358), and Kaul (1974, 85).
2. For some interesting discussions regarding the bases of power see Crabb (1965), Morgenthau (1973), Organski (1968), and Stoessinger (1969).
3. Regions were delineated in accordance with the Singer-Russett-Small nation number system. See Singer and Small (1972).
4. See Dhar (1957, 5), Gowing (1974, 63, 184, 220, 407), Groom (1974, 44), Joshua and Hahn (1973, 11), Kohl (1971, 98, 358), Mendl (1969, 169), Scheinman (1965, 188-95), and Subramanyam (1974, 122, 127, 134-35).
5. See, for example, Cox and Jacobson (1974), Modelski (1974), Organski (1968), Singer, Bremer, and Stuckey (1972), Singer and Small (1972), and Spiegel (1972).
6. This list was derived from a survey of historians and political scientists conducted by the University of Michigan's Correlates of War Project.
7. Except for Cuba, Brazil was the first Latin American nation to recognize the new Angolan regime. See *New York Times* (1976a, 3).
8. *New York Times* (1976b, 1). It has been reported that Brazilian officials had in fact been seeking such "recognition" since 1974. See Gall (1976).
9. The identification of nuclear weapons countries was, of course, a by-product of the data collection described in chapter 3. The data on military alliances were obtained from the University of Michigan Correlates of War Project and supplemented with data from the International Institute for Strategic Studies (series b) and *Keesing's Treaties and Alliances of the World* (1976).
10. The dispute data were drawn primarily from Butterworth (1977) and cross-checked against Kende (1971), Singer and Small (1972), and Wood (1968).
11. The definition of war used in this study uses the criterion employed by the University of Michigan's Correlates of War Project—that is, 1000 battle deaths. See Singer and Small (1972).
12. The choice of the five- and ten-year effect periods was subjective, but they seem like reasonable estimates to begin with.

13. Of course these groups are not meant to be mutually exclusive—a nation may well appear in more than one category. Britain, for instance, is in the global power group, the regional power group, and the alliance partner group.
14. In general, most of the "targeted" alliances were cold war alliances. The Organization of American States, for instance, has no specific target.
15. Declared neutrals should not be confused with nonaligned nations. The former are legally bound not to ally with *any* country; the latter may indeed form alliances—outside the East/West rivalry. See Holsti (1972, chap. 4).
16. A regionally allied global power is simply a global power that has a defense pact with any nation in the region. Eventually, I would like to include "unattached" military bases as well—for example, Diego Garcia with respect to the United States.
17. An obvious problem is the difference in costs for weaponry between countries. However, as a first-cut approach the method has a certain intuitive appeal. In one sense the international arms trade has in many ways acted as a cost equalizer between the particular nation pairs most likely to be engaged in a conflict. Moreover, the ratio threshold employed to indicate conventional superiority is high enough to overcome much of the "noise" created by differing military pay scales, factor costs, and such, between countries.
18. For those cases in which a single nation was the target of an alliance, the alliance members' military capabilities were summed before computing R_TP.
19. The empirical findings of Sabrosky (1975) appear to support these three-to-one formulations.
20. The data were taken from Banks (1975). The choice of these particular indicators was based on the research findings of Rummel (1972) and Wilkenfeld (1972).
21. The data were obtained from the University of Michigan's Correlates of War Project.
22. Makaizumi (1966, 83), Spence (1974, 733), Sommer (1966, 51), Stockholm International Peace Research Institute (1972, 71), Birnbaum (1966, 73-74), and Quester (1973, 139).
23. This is one of two factors hypothesized to play a dual role in the nuclear proliferation literature, so it appears in both the motive condition and dissuasive condition lists.
24. Data on coups and revolutions were taken from Banks (1975). The choice of these particular indicators was based on the research finding of Rummel (1972), and Wilkenfeld (1972).

CHAPTER 4

1. A nation-year case represents an observation of a single country in a single year. Annual data observations of Japan from 1970 to 1979 reflect ten nation-year cases. Nation-year observations for countries that have latent capacities but have not made proliferation decisions run from the date of first acquiring a latent capacity through 1980. Nation-year observations for countries that have made proliferation decisions run from the year of the initiating decision through the year in which the first weaponry is produced or until the program is canceled.

CHAPTER 5

1. For some interesting discussions of the differences between classical probability theory and Bayesian probability theory see Winkler (1972), Savage (1972), and Schmitt (1969).
2. Besides the sources listed in tables 3 and 4, see Rosecrance (1966), U.S. Senate (1976), Yaeger (1980), and Lefever (1979).
3. Particularly good discussions of this one-to-one correspondence, or one-to-several correspondence, can be found in Dunn and Kahn (1976), Greenwood (1977), Epstein (1976), Office of Technology Assessment (1977), Yaeger (1980), Dunn (1982), and Potter (1982).
4. Cross-tabulations of the motive conditions were all consistent with the statistical independence assumption necessary for this estimation procedure.
5. For those who are interested, an analogous test of the technological imperative hypothesis—predicting proliferation decisions whenever a latent capacity is present—produces phi $= 0.23$; lambda $= 0$; and tau $= 0.05$. These results argue very strongly that the data are inconsistent with the technological imperative hypothesis.
6. A statistical test for systematic error asks: In sampling six cases out of sixty, how likely are we to draw four of a single nation that comprises 10 percent of the sixty cases? The answer is about 0.0005. When the four cases are also in time order, the probability drops further.
7. Wohlstetter (1979, 43) argues that Egypt is a moderate infrastructure country.

CHAPTER 6

1. This notion of dynamic convergence is developed in Rose-crance (1966) and, most explicitly, in Nye (1977).
2. For more complete examinations of South Africa's possible interest in nuclear weapons see Spence (1974), Adelman and Knight (1979), and Betts (1980).
3. The extent of India's involvement in nuclear science and engineering was much more substantial than portrayed here. Its researchers were busily involved in heavy water production, special reactor designs, power research, and so forth.
4. Historical studies clearly show that Shastri's approval of a nuclear explosives project was directly related to security issues linked to the 1962 war loss and the 1964 Chinese nuclear test.
5. Note that the nuclear propensity model used here did not explicitly include a variable to represent the kind of asymmetric dependency described in the South Korean case. There is certainly no reason it could not be added to the motivational profile as a sixteenth variable. It would probably be most applicable to just a few states such as Israel, Taiwan, and South Korea.
6. There were also plans to provide Iraq with hot cells—laboratory facilities where small-scale plutonium reprocessing is possible.
7. One source reports that the Iraqis initially requested a high power natural uranium fueled, graphite moderated reactor, a request the French turned down. If this is true, it would have been an optimum system for plutonium production for weapons purposes. See Winkler (1981).

APPENDIX A

1. For Germany see Brown and McDonald (1977, part 3) and Irving (1967); for Japan see Shapely (1978) and Weiner (1978).
2. The most extensively researched survey of this literature is by Holloway (1980). See also Wolfe (1970, 35).
3. See Dhar (1957, 77, 197), Scheinman (1965), and Zoppo (1964). See also Kelly (1960) and Kohl (1971).
4. The relative necessity of testing is discussed in detail in chapter 2.
5. Excellent summaries are provided by Dowty (1978), Harkavy (1977b), and Haselkorn (1974).

6. The fundamental relation between an atomic bomb and a "peaceful" nuclear explosive is discussed in chapter 2.
7. Kaul (1974, 29), Marwah (1977, 101), Bray and Moodie (1977, 114-15), Kapur (1976), Paloose (1978).
8. This section is based on reports in the *Washington Post* (1978) and the *Los Angeles Times* (1978) and on several interviews I conducted.

Appendix B

1. See Dean (1953, 44-50) for a brief, nontechnical discussion. See also Glasstone and Sesonske (1963, 455-61).
2. Van Cleave (1974, 39). Moreover, very detailed instructions for uranium processing can be found in the open literature.
3. The required aluminum is of general commercial grade (2S quality) and is therefore openly available for import. For a discussion of various "canning" alternatives, see Wilkinson (1962).
4. Lamarsh (1976, 12) estimates about six months. However, even though the canning is relatively easy, this seems a very optimistic estimate. Practice suggests that something closer to a year is more appropriate.
5. This distinction is noted in Office of Technology Assessment (1977, 177-79) and U.S. House of Representatives (1975, 94-99).
6. Office of Technology Assessment (1977, 177). For a discussion of hot cells see Beck (1957, chap. 8). See also Jackson and Sadowski (1955) and Ohlgren et al. (1955).
7. This was in fact the procedure used at Hanford in the United States Manhattan Project. See Bebbington (1976, 30).
8. The determination of the value for Y is discussed below.
9. At its highest point, during the mid-1950s, the United States atomic weapons program was consuming 10 percent of the electrical power generated in the United States.
10. Again, the problems inherent in the production of "nuclear-grade" graphite are not so much problems of process engineering as problems of acquiring raw material feed of high purity. Obviously, the higher the purity of the input material, the less effort is required to remove impurities during processing. See Glasstone and Sesonske (1963, 439), Legendre (1955), Nightingale (1962), and Currie, Hamister, and MacPherson (1955, chap. 3).
11. Though a higher correlation coefficient could have been

obtained using a noninteger value for Y, I deemed this an unnecessary concession to "false" precision. The estimated accuracy of the indicators is probably about + 2 years. Thus, recording fractions of a year would add no additional information.

Appendix C

1. See Clegg and Foley (1958, chap. 2). They note that the cost of uranium mining could vary from $800,000 to $1.4 million (1958 dollars). Mabile (1958, 40-43) and Andriot and Gaussens (1958) peg an equivalent French operation at over $2 million (1958 dollars).
2. *Nucleonics*, April 1960, 38.
3. *Business Week*, 21 April 1956, 31.
4. Office of European Economic Cooperation (1956, annex vii); 0.35 scaling.
5. This turned out to be 0.35 over a wide range of low-capacity plants.
6. Murray et al. (1958); stainless steel cladding is used. Andriot and Gaussens (1958) note that the capital costs associated with fabrication plants for U-metal are invariant with respect to the type of cladding used (zirconium, stainless steel, aluminum, etc.). United Nations (1968) uses an implicit 0.5 scaling in its analysis.
7. Patton, Googin, and Griffith (1963, chap. 7). This source also notes that the construction for nuclear plants in India typically costs 30 percent less than comparable facilities in the United States. Thus, when we inflate the Indian cost figure by the appropriate amount, it produces a scaled cost for our hypothetical plant of some $2.0 million. See also Kaul (1974).
8. The exponent is 0.35. Since the capital investment required for low burnup plants is not so much a function of the process equipment as of their absolute size, the scale factor tends to hover between 0.25 and 0.40; see Bosch, Oszusky, and Asyee (1968), Little (1968, 221), Schwennesen (1957).
9. Owing to the low burnup of the fuel, the waste product could simply be "bottled and buried" in stainless steel/concrete containers, as was originally done at Hanford (United States), Oak Ridge (United States), Windscale (United Kingdom), and Marcoule (France).
10. United Nations (1968, 55). U. S. Atomic Energy Commission

(1971, 85) calculates that a plant to produce plutonium oxide from PU ($^{NO}3$) would cost about $500,000 (1960 dollars).

11. An exponential scale factor of 0.35 was used in "scaling up" the Indian Trombay plant.

12. Even before 1960, a large number of individuals from more than forty-five countries participated in training programs sponsored by the United States Atomic Energy Commission. See Joint Committee on Atomic Energy (1960, 245-98).

13. Computed using Beaton (1966a, 34-35), Beaton and Maddox (1962, 93), Dean (1953, 265-66), Dhar (1957, 148), Hohenemser (1962, 240), Scheinman (1965, 181-84, and United Nations (1968, 62).

14. Seshagiri (1975, 69), United Nations (1968, annex IV). This is also consistent with estimated operating costs using the FUELCO model; see U. S. Atomic Energy Commission (1971), as well as International Atomic Energy Agency (1968).

Bibliography

Adelman, Kennet L., and Albion W. Knight. 1979. Can South Africa go nuclear? *Orbis* 23, no. 3 (fall):633-48.

Akimov, V., and A. Pamor. 1973. Military-economic preparations of the Chinese People's Republic. *Voyennaya Mysl'* 9:99-108.

Allison, Graham T. 1979. *Essence of decision*. Boston: Little, Brown.

Andriot, J., and J. Gaussens. 1958. Fuel cycles, the production of uranium, power plant programs and relevant investments. In *Peaceful uses of atomic energy*, Vol. 2. P/1197, 165-83. Geneva: United Nations.

Ashbrook, Arthur G. 1975. China economic overview, 1975. In Joint Economic Committee 1975.

Bader, William. 1968. *The United States and the spread of nuclear weapons*. New York: Pegasus.

Banks, Arthur. 1971. *Cross-polity time series*. Cambridge: MIT Press.

———. 1975. *Cross-national time series: 1815-1973*. Ann Arbor, Mich.: Inter-University Consortium for Political and Social Research.

Barber, Richard J., Associates, Inc. 1975. *LDC nuclear power prospects, 1975-1990: Commercial, economic and security implications*. Washington, D.C.: Barber Associates.

Barnaby, C. F. 1969a. *Preventing the spread of nuclear weapons*. New York: Humanities Press.

Baxter, James P. 1946. *Scientists against time*. Cambridge: MIT Press.

Beaton, Leonard. 1966a. Capabilities of non-nuclear powers. In Buchan 1966, 13-38.

———. 1966b. *Must the bomb spread*. Baltimore: Penguin Books.

Beaton, Leonard, and John Maddox. 1962. *The spread of nuclear weapons*. New York: Praeger.

Bebbington, William P. 1976. The reprocessing of nuclear fuels. *Scientific American*, December 30-41.

Beck, Clifford K. 1957. *Nuclear reactors for research*. Princeton, N.J.: Van Nostrand.

Belden, Thomas G. 1977. Indications, warning and crisis operation. *International Studies Quarterly* 21, no. 1 (March): 181-98.

Betts, Richard. 1977. Paranoids, pigmies, pariahs, and non-proliferation. *Foreign Policy*, no. 26 (Spring): 157-83.

———. 1980. South Africa. In Yaeger 1980, 283-308.

Bhargava, G. S. 1976. *India's security in the 1980's*. London: International Institute for Strategic Studies.

Binder, David. 1976. U. S. fears spread of A-arms in Asia. *New York Times*, 31 August, 6.

Birnbaum, Karl. 1966. The Swedish experience. In Buchan 1966, 68-75.

Blalock, Hubert M., Jr. 1960. *Social*

217

statistics. New York: McGraw-Hill.

Bosch, T., F. Oszusky, and J. Asyee. 1968. Economic aspects of reprocessing nuclear fuel. In *Economics of nuclear fuels*, Vienna: International Atomic Energy Agency. 327-38.

Bray, Frank T. J., and Michael L. Moodie. 1977. Nuclear politics in India. *Survival* 20, no. 3 (May/June): 111-16.

Brennan, Donald. 1976. A comprehensive test ban: Everybody or nobody. *International Security* 1, no. 1 (Summer): 92-117.

Brown, Anthony C., and Charles B. McDonald. 1977. *The secret history of the atomic bomb*. New York: Dial Press/James Wade.

Brown, Seyom. 1974. *New forces in world politics*. Washington, D.C.: Brookings Institution.

Buchan, Alastair, ed. 1966. *A world of nuclear powers*. Englewood Cliffs, N.J.: Prentice-Hall.

Bucy, J. Fred. 1977. On strategic technology transfers to the Soviet Union. *International Security* 1, no. 4 (Spring): 25-43.

Business Week. 1955. Uranium: Now it's a race for production. 5 February, 66-75.

———. 1956. Uranium industry leaps to maturity. 21 April, 30-32.

Butterworth, Robert L. and Scranton, Margaret E. 1976. *Managing interstate conflict, 1945-1974: Data with synopses*. Pittsburgh: University Center for International Studies, University of Pittsburgh.

Campbell, Donald T., and Julian C. Stanley. 1963. *Experimental and quasi experimental designs for research*. Chicago: Rand McNally.

Caporaso, James A., and Leslie L. Roos, Jr. 1973. *Quasi-experimental approaches*. Evanston, Ill., Northwestern University Press.

Carlton, D. 1966. The problem of civil aviation in British air disarmament policy, 1919-1934. *Royal United Service Institution Journal* 3, no. 644.

———. 1969. Verification and security guarantees. In Barnaby 1969a, 127-43.

Chan, Steven. 1980. Incentives for nuclear proliferation: The case of international pariahs. *Journal of Strategic Studies* 3, 1, (May):26-43.

Chastain, Joel W., Jr., ed. 1958. *U. S. research reactor operation and use*. Reading, Mass.: Addison-Wesley.

Clegg, John, and Dennis Foley. 1958. *Uranium ore processing*. Reading, Mass: Addison-Wesley.

Cooley, John, and Lisa Kaufman. 1980. Deterring the Qaddafi bomb. *Washington Post*, 23 December, 15.

Cortney, William. 1980. Brazil and Argentina. In Yaeger 1980, chap. 11.

Cox, Robert, and Harold Jacobson. 1974. *The anatomy of influence*. New Haven: Yale University Press.

Crabb, Cecil V. 1965. *American foreign policy in the nuclear age*. New York: Harper and Row.

Culler, Floyd. 1963. *The effect of scale up on fuel cycle costs for enriched fuel and natural uranium fuel systems*. ORNL-TM-564 Oak Ridge, Tenn.: Oak Ridge National Laboratory.

Currie, L. M., V. C. Hamister, and H. G. MacPherson. 1955. *The production and properties of graphite for reactors*. National Carbon Company.

Cyert, R., and J. March. 1963. *A behavioral theory of the firm*. Englewood Cliffs, N. J.: Prentice-Hall.

Dawson, F. G., D. E. Deonigi, and E. A. Eschbach. 1965. Plutonium buildup and depletion. *Nucleonics* 23, no. 8 (August):101-5.

Dean, Gordon. 1953. *Report on the atom*. New York: Alfred Knopf.

DeRivera, Joseph. 1968. *The psychological dimension of foreign policy*. Columbus, Ohio: Merrill.

Deutsch, Karl. 1963. *The nerves of government*. New York: Free Press.

DeWeerd, H. A. 1964. British-American collaboration on the A-bomb in World War II. In Rosecrance 1964a, 29-47.

Dhar, Sailendra Nath. 1957. *Atomic weapons in world politics*. Calcutta: Das Gupta.

Dietz, David. 1954. *Atomic science,*

bombs and power. New York: Dodd, Mead.

Dowty, Alan. 1972. Israeli perspectives on nuclear proliferation. In Holst 1972a, 142-51.

———. 1978. Nuclear proliferation: The Israeli case. *International Studies Quarterly* 22, 1 (March):79-120.

Dunn, Lewis A. 1982. *Controlling the bomb*. New Haven: Yale University Press.

Dunn, Lewis A., and Herman Kahn. 1976. *Trends in nuclear proliferation, 1975-1995*. Croton-on-Hudson, N. Y.: Hudson Institute.

Eatherly, W. P., and E. L. Piper. 1962. Manufacture. In Nightingale 1962, 22-50.

Emelyanov, V. S. 1969. Nuclear reactors will spread. In Barnaby 1969a, 65-71.

Enthoven, Alain C., and Wayne Smith. 1971. *How much is enough?* New York: Harper and Row.

Epstein, William. 1976. *The last chance*. New York: Free Press.

———. 1977. Why states go—and don't go—nuclear. *Annals of the American Academy of Political and Social Science*, 430 (March):16-28.

Faulstich, H. 1969. The problem of safeguards with special reference to the GDR. In Barnaby 1969a, 163-66.

Fisher, Adrian S. 1967. Issues involved in a non-proliferation agreement. In Kertesz 1967, 36-52.

Ford-Mitre Nuclear Energy Policy Study Group. 1977. *Nuclear power issues and choices*. Cambridge, Mass.: Ballinger.

Frejacques, C., and R. Galley. 1964. Enseignements tires des etudes et realisations francaises relas a la separation des isotopes de l'"Uranium." In *Peaceful uses of atomic energy*, vol. 12. Geneva: United Nations.

Gall, Norman. 1976. Atoms for Brazil, dangers for all. *Foreign Policy*, no. 23 (Summer):155-201.

Gallois, Pierre. 1961. *The balance of terror: Strategy for the nuclear age*. Boston: Houghton Mifflin.

Garris, Jerome. 1972. Sweden's debate on the proliferation of nuclear weapons. A paper of the Southern California Arms Control and Foreign Policy Seminar.

Garthoff, Raymond. 1978. On estimating and imputing intentions. *International Security* 2, no. 3 (Winter):22-31.

Gelber, Harry G. 1972. Australia and nuclear weapons. In Holst 1972a, 100-120.

———. 1973. *Nuclear weapons and chinese policy*. Adelphi Paper no. 99. London: International Institute for Strategic Studies.

Geneste, Marc. 1976. The nuclear land battle. *Strategic Review* 4 (Winter):329-33.

Glasstone, Samuel. 1955. *The effects of atomic weapons*. Washington, D. C.: U. S. Government Printing Office.

———. (1962). *The effects of nuclear weapons*. Washington, D.C.: U. S. Government Printing Office.

Glasstone, Samuel, and Alexander Sesonske. 1963. *Nuclear reactor engineering*. New York: Van Nostrand Reinhold.

Gorbachev, A. 1981. Peking's nuclear ambition. *Krasnaya Zvezda*, 25 January, 3.

Goldschmidt, Bertrand. 1962. The French atomic energy program. *Bulletin of the Atomic Scientists* 18, no. 7 (September):39.

———. 1977. A historical survey of non-proliferation policies. *International Security* 2, no. 1 (Summer):69-87.

Gowing, Margaret. 1964. *Britain and atomic energy*. New York: St. Martin's.

———. 1974. *Independence and deterrence, Britain and atomic energy, 1945-1952*. New York: Macmillan.

Greenwood, Ted, Harold A. Feiveson, and Theodore B. Taylor. 1977. *Nuclear proliferation*. New York: McGraw-Hill.

Greenwood, Ted, George W. Rathjens, and Jack Ruina. 1976. *Nuclear proliferation and weapons proliferation*.

London: International Institute for Strategic Studies.

Groom, A. J. R. 1974. *British thinking about nuclear weapons*. London: Frances Pinter.

Groves, Leslie. 1962. *Now it can be told*. New York: Harper.

Gupta, Sisir. 1966. The Indian dilemma. In Buchan 1966, 55-67.

Guthrie, C. E. 1957. *The effect of the radiochemical reprocessing industry's growth on spent fuel reprocessing costs*. Oak Ridge National Laboratory 2279. Oak Ridge, Tenn.

Haagerup, Niels J. 1972. Nuclear weapons and Danish security policy. In Holst 1972a, 34-41.

Hall, D. B. 1972. "The adaptability of fissile materials to nuclear explosives. In Leachman and Althoff 1972, 275-83.

Halperin, Morton. 1965. *China and the bomb*. New York: Praeger.

————. 1974. *Bureaucratic politics and foreign policy*. Washington, D.C.: Brookings.

Harkabi, Y. 1966. *Nuclear war and nuclear peace*. Jerusalem: Israel Program for Scientific Translations.

Harkavy, Robert E. 1977a. The pariah state syndrome. *Orbis* 21, no. 3 (Fall):623-50.

————. 1977b. Spectre of a Middle East holocaust. Mimeograph. Boulder: Graduate School of International Studies, University of Colorado.

Haselkorn, Avigator. 1974. Israel: An option to a bomb in the basement. In Lawrence and Larus 1974, 149-82.

Higinbothom, W. A., and Jo Pomerance. 1969. Non-proliferation and the arms race. In Barnaby 1969a, 192-202.

Hitch, Charles J., and Roland McKean. 1967. *The economics of defense in the nuclear age*. Cambridge: Harvard University Press.

Hoglund, R. L., J. Schachter, and E. von Halle. 1965. Diffusion separation methods. In *Encyclopedia of chemical technology*. New York: John Wiley.

Hohenemser, Christoph. 1962. The nth

country problem today. In *Disarmament: Its politics and economics*, ed. Seymour Melman, 238-78. Boston: American Academy of Arts and Sciences.

Holloway, David. 1980. Entering the nuclear arms race: The Soviet decision to build the atomic bomb, 1939-1948. Working Paper no. 9, Washington, D. C.: Wilson Center.

Holst, Johan J. 1972a. *Security, order and the bomb*. Oslo: Universitetsforlaget.

————. 1972b. The nuclear genie: Norwegian politics and perspectives. In Holst 1972a, 42-60.

Holsti, K. J. 1972. *International politics: A framework for analysis*. Englewood Cliffs, N. J.: Prentice-Hall.

Horner, Charles. 1973. The production of nuclear weapons. In *The military and political power in China in the 1970s*, ed. William Whitson. New York: Praeger.

Hsieh, Alice Langley. 1964a. Communist China and Nuclear Force. In Rosecrance 1964a, 157-85.

————. 1964b. The Sino-Soviet nuclear dialogue: 1963. *Journal of Conflict Resolution* 8, no. 2 (June):99-115.

Huber, A. P. et al. 1958. Multiton production of flourine for the manufacture of UF6. In *Peaceful uses of atomic energy*, vol. 2, 172-80. Geneva: United Nations.

Imai, Ryukichi. 1972. The changing role of nuclear technology in the post-NPT world: A Japanese view. In Holst 1972a, 120-30.

India, Department of Energy. 1965. *Ten years of atomic energy in India*. Bombay: Department of Energy.

International Atomic Energy Agency. 1959. *Directory of nuclear reactors*. Vienna: IAEA.

————. 1968. *Economics of nuclear fuels*. Vienna: IAEA.

————. 1969. *Power and research reactors in member states* Vienna: IAEA.

————. *Power reactors in member states*. Series. Vienna: IAEA.

International Institute for Strategic Studies. 1977. *The military balance,*

1977-1978. London: IISS.

————. 1981. *The military balance, 1980-1981.* London: IISS

Irving, David. 1967. *The German atomic bomb.* New York: Simon and Schuster

Jackson, H. K., and G. S. Sadowski. 1955). Design and operation for direct maintenance fuel separation. *Nucleonics* 13, no. 8 (August):22-25.

Jacobson, Harold, and Eric Stein. 1966. *Diplomats, scientists, and politicians: The U. S. and the nuclear test ban negotiations.* Ann Arbor: University of Michigan Press

Janis, Irving L. 1972. *Victims of groupthink.* Boston: Houghton Mifflin.

Jensen, Lloyd. 1974. *Return from the nuclear brink.* Lexington, Mass.: Lexington Books.

Jevis, Robert. 1976. *Perception and misperception in international politics.* Princeton: Princeton University Press.

Joint Committee on Atomic Energy. 1960. *Report.* Vols. 1-5. Washington, D. C.: U. S. Government Printing Office.

Joint Economic Committee. 1975. China: A reassessment of the economy. Committee Print, 94th Congress, 1st session, 10 July.

Joshua, Wynfred, and Walter F. Hahn. 1973. *Nuclear politics: America, France, and Britain.* Beverly Hills, Calif.: Sage Publications.

Kapur, Ashok. 1976. *India's nuclear option: Atomic diplomacy and decision making.* New York: Praeger.

Kato, Saburo. 1974. Japan: Quest for strategic compatability. In Lawrence and Larus 1974, 183-204.

Kaul, Ravi. 1974. *India's nuclear spin-off.* Allahabad, India: Onanakya.

Keesing's treaties and alliances of the world. 1976. New York: Scribner's.

Kelly, George A. 1960. The political background of the French A-bomb. *Orbis* 4, no. 3 (Fall):284-306.

Kemp, Geoffrey. 1974. *Nuclear forces for medium powers.* London: International Institute for Strategic Studies.

Kende, Istvan. 1971. Twenty-five years of local wars. *Journal of Peace Research*, no. 1:5-22.

Kertesz, Stephen D. 1967. *Nuclear non-proliferation in a world of nuclear powers.* Notre Dame: University of Notre Dame Press.

Khrushchev, N. 1974. *Khrushchev remembers: The last testament.* Boston: Little, Brown.

Kirk, Raymond E., and Donald F. Othner. 1949. *Encyclopedia of chemical technology.* New York: Interscience Encyclopedia.

Kissinger, Henry A. 1957. *Nuclear weapons and foreign policy.* New York: Harper.

————. 1964. *A world restored.* New York: Grosset and Dunlap.

Kohl, Wilfrid L. 1971. *French nuclear diplomacy.* Princeton: Princeton University Press.

Kramish, Arnold. 1959. *Atomic energy in the Soviet Union.* Stanford: Stanford University Press.

————. 1963. *The peaceful atom in foreign policy.* New York: Harper and Row.

————. 1964. The emergent genie. In Rosecrance 1964a, chap. 9.

Lagrange, Pierre, and Olegh Bilous. 1958. Re-enrichment of depleted uranium by passage through a gaseous diffusion installation. In *Peaceful uses of atomic energy.* Geneva: United Nations.

Lamarsh, John R. 1976. On the construction of plutonium producing reactors by small and/or developing nations. Prepared for the Congressional Research Service of the Library of Congress. *Export reorganization act of 1976*, 1326-55. Hearings before the Senate Committee on Government Operations, 94th Congress, 2d session, 19, 20, 29, 30 January and 9 March.

Lane, James A. 1957. An evaluation of Geneva and post-Geneva nuclear power economic data. In *The economics of nuclear power*, ed. J. Gueron et al. New York: McGraw-Hill.

Lapp, Ralph. 1968. *The weapons cul-*

ture. Baltimore: Penguin Books.

Laurence, William L. 1959. *Men and atoms*. New York: Simon and Schuster.

Lawrence, Robert M., and Joel Larus. 1974. *Nuclear proliferation phase II*. Lawrence, Kans.: Allen Press.

Leachman, Robert B., and Phillip Althoff. 1972. *Preventing nuclear theft: Guidelines for industry and government*. New York: Praeger.

Lefever, Ernest W. 1979. *Nuclear arms in the Third World*. Washington, D. C.: Brookings Institution.

Legendre, P. 1955. The production of nuclear graphite in France. In *Proceedings of the International Conference on the Peaceful Uses of Atomic Energy*. Vol. 2. New York: United Nations.

Leurdijk, J. Henk. 1972. Nuclear weapons in Dutch foreign policy. In Holst 1972a, 19-33.

Lewis, W. H. 1955. *The ORNL metal recovery plant*. ORO 144, 49-69. (Oak Ridge, Tenn.: Oak Ridge National Laboratory.

Liddell Hart, B. H. 1954. *Strategy*. New York: Praeger.

Little, Arthur D., Inc. 1968. *Competition in the nuclear power supply industry*. NYO 3853-1 TID UC-2. Washington, D. C.: U. S. Government Printing Office.

Los Angeles Times. 1977. Low-grade fuel used in A-bomb test. 14 September.

Mabile, J. 1958. Development of the uranium mining industry in France and the French Union. In *Peaceful uses of atomic energy*, vol. 2. Geneva: United Nations.

McPhee, John. 1974. *The curve of binding energy*. New York: Farrar, Straus, and Giroux.

Makauzumi, Kei. 1966. The problem for Japan. In Buchan 1966, 76-88.

Maloney, James O., et al. 1955. *Production of heavy water*. New York: McGraw-Hill.

Martenson, M. 1968. Economics of uranium enrichment by gaseous diffusion. In *Economics of nuclear fuels*, 275-295. Vienna: International Atomic Energy Agency.

Martin, Wayne R. 1977. Measure of international military commitments for crisis early warning. *International Studies Quarterly* 21, no. 1 (March):151-78.

Marwah, Onkar. 1977. India's nuclear and space programs: Intent and policy. *International Security* 2, no. 2 (Fall):96-121.

Maxwell, A. E. 1961. *Analyzing qualitative data*. London: Chapman and Hall.

Mendl, Wolf. 1969. The spread of nuclear weapons: Lessons from the past. In Barnaby 1969a, 169-79.

Meyer, Stephen M. 1976. Some possibilities for the estimation of constrained parameters in regression analysis. Ann Arbor: University of Michigan Department of Political Science. Mimeographed.

———. 1978a. A cost-capacity analysis of uranium milling plants. Ann Arbor: University of Michigan Department of Political Science. Mimeographed.

———. 1978b. A cost-capacity analysis of gaseous diffusion plants. Ann Arbor: University of Michigan Department of Political Science. Mimeographed.

———. 1978c. A cost-capacity analysis of plutonium reprocessing plants for low-burnup fuels. Ann Arbor: University of Michigan Department of Political Science. Mimeographed.

———. 1978d. Probing the causes of nuclear proliferation. Ph.D. diss., University of Michigan.

———. 1981b. Nuclear decisionmaking in India. Cambridge: Center for International Studies, MIT. Mimeographed.

Millar, T. B. 1974. Australia: Recent ratification. In Lawrence and Larus 1974, 69-85.

Modelski, George. 1959. *Atomic energy in the Communist bloc*. New York: Cambridge University Press.

———. 1974. *World power concentrations*. Morristown, N. J.: General

Learning Press.

Morgenthau, Hans. 1973. *Politics among nations*. New York: Alfred A. Knopf.

Murphy, Charles H. 1972. Mainland China's evolving nuclear deterrent. *Bulletin of the Atomic Scientists* 28 (January):28-35.

Murray, J. P., F. S. Patton, R. F. Hibbs, and W. L. Griffith. 1958. Economics of unirradiated processing phases of uranium fuel cycles. In *Peaceful uses of atomic energy*. Vol. 13. P/439, 582-601. Geneva: United Nations.

Nacht, Michael. 1981. The future unlike the past: Nuclear proliferation and American security policy. In Quester 1981, 193-212.

National Planning Association. 1960. *The nth country problem and arms control*. Washington, D. C.: NPA.

Nerlich, Uwe. 1972. Nuclear weapons and European politics: Some structural interdependencies." In Holst 1972a, 74-92.

Nerlich, Uwe. 1974. The Federal Republic of Germany: Constraining the inactive. In Lawrence and Larus 1974, 86-111.

New York Times. 1976a. Brazil and U. S. to consult regularly, Kissinger announces on Brasilia visit. 20 February, 3.

———. 1976b. U. S. and Brazil sign accord on ties. 22 February, 1.

———. 1977a. Statement by Soviets says South Africa nears A-bomb test. 9 August, 9.

———. 1977b. South Africa says it is not planning atomic bomb tests. 22 August, 3.

———. 1977c. France says data show South Africa plans atomic test. 23 August, 1.

———. 1977d. U. S.-Soviet exchange about South Africa said to improve ties. 29 August, 1.

———. 1977e. South Africa stirs new A-arms flurry. 31 August, 2.

Nightingale, Richard E. ed. 1962. *Nuclear graphite*. New York: Academic Press.

Nonproliferation Alternative Systems Assessment Program (NASAP). 1980. *Nuclear proliferation and civilian nuclear power*. Washington, D. C.: U. S. Department of Energy.

Nye, Joseph. 1977. Time to plan for the next generation of nuclear technology. *Bulletin of the Atomic Scientists* 33, no. 8 (October):38-41.

Office of Technology Assessment. 1977. *Nuclear proliferation and safeguards*. Washington, D. C.: National Technical Information Service.

Ohlgren, H. A., et al. 1955. Reprocessing reactor fuels. *Nucleonics* 13, no. 3 (March):18-21.

Olgaard, P. L. 1969. The Soviet-American draft non-proliferation treaty: Will it work? In Barnaby 1969a, 213-28).

Olsen, J. L., E. C. Savage, J. H. Tucker, and R. L. Beck. 1958. The economic environment for nuclear power in Canada. In *Peaceful uses of atomic energy*. Vol. 13. P/207, 624-33. Geneva: United Nations.

Organization for European Economic Cooperation. 1956. *Nuclear energy*. Paris: OEEC.

Organski, A. F. K. 1968. *World politics*. New York: Alfred A. Knopf.

Paloose, T. T. 1978. *Perspectives on India's nuclear policy*. New Delhi: Young Asia Publications.

Palumbo, Dennis. 1969. *Statistics in political and behavioral science*. New York: Appleton-Century-Croft.

Patton, Finis S., John M. Googin, and William L. Griffith. 1963. *Enriched uranium processing*. New York: Pergamon Press.

Perlmutter, Amos. 1982. The Israeli raid on Iraq. *Strategic Review* 10, 1 (Winter):34-43.

Pierre, Andrew J. 1972. *Nuclear politics*. London: Oxford University Press.

Planning Research Corporation. 1968. *Multivariate analysis of combat*. Los Angeles: PRC.

Potter, William. 1982. *Nuclear power and nonproliferation*. Cambridge,

Mass.: Oeleschlager, Gunn, and Hain.

Prakash, B., and N. K. Rao. 1961. Fabrication of fuel elements at Trombay, India. In *Fuel element fabrication.* New York: Academic Press.

Prawitz, J. 1969. Safeguards and related arms control procedures. In Barnaby 1969a, 113-26.

———. 1972. Sweden: A non-nuclear weapon state. In Holst 1972a, 61-73.

Quester, George. 1973. *The politics of nuclear proliferation.* Baltimore: Johns Hopkins University Press.

———. ed. 1981. *Nuclear proliferation: Breaking the chain.* Madison: University of Wisconsin Press.

Reynolds, H. T. 1977. *Analysis of nominal data.* Beverly Hills, Calif.: Sage Publications.

Roberts, Chalmers. 1970. *The nuclear years.* New York: Praeger.

Rosecrance, Richard, ed. 1964a. *The dispersion of nuclear weapons strategy and politics.* New York: Columbia University Press.

———. 1964b. British incentives to become a nuclear power. In Rosecrance 1964a, 48-65.

———. 1964c. British defense strategy: 1945-1952. In Rosecrance 1964a, 66-86.

———. 1966. *Problems of nuclear proliferation.* UCLA Security Studies Project no. 7. Los Angeles: University of California.

Rowen, Henry, and Richard Brody. 1980. The Middle East. In Yaeger 1980, 177-240.

Rummel, R. J. 1972. *The dimensions of nations.* Beverly Hills, Calif.: Sage Publications.

Russett, Bruce M. 1970. *What price vigilance?* New Haven: Yale University Press.

Sabrosky, Alan. 1975. The war-time reliability of interstate alliances, 1816-1965. A paper delivered at the Sixteenth Annual Conference of the International Studies Association, Washington, D.C., 19-22 February.

Salmon, R., et al. 1972. Price forecasting and resource utilization for the fuel-cycle industry of the USA. In *Peaceful Uses of Atomic Energy,* vol. 4. New York: United Nations.

Sanders, Benjamin. 1975. *Safeguards against nuclear proliferation.* Cambridge, Mass.: MIT Press.

Savage, Leonard. 1972. *The foundations of statistics.* New York: Dover.

Scheinman, Lawrence. 1965. *Atomic energy policy in France under the fourth republic.* Princeton: Princeton University Press.

Schleslinger, James R. 1967. Nuclear spread: The setting of the problem. In Kertesz 1967, 8-28.

Schmitt, Samuel A. 1969. *Measuring uncertainty.* Reading, Mass.: Addison-Wesley.

Schwennesen, J. L. 1957. Capital and operating cost information on several existing U. S. nuclear fuel reprocessing plants. In *Symposium on the reprocessing of irradiated fuels.* Washington, D. C.: U. S. Atomic Energy Commission, Book 3.

Seshagiri, Narasimhiah. 1975. *The bomb! Fallout of India's nuclear explosion.* Delhi: Vikas.

Shapley, Deborah. 1978. Nuclear weapons history: Japan's wartime bomb projects revealed. *Science* 199 (13 January):152-57.

Singer, David, Stuart Bremer, and John Stuckey. 1972. Capability distribution, uncertainty, and major power war, 1820-1965. In *Peace, war and numbers,* ed. Bruce Russet. Beverly Hills: Sage.

Singer, J. David, and Melvin Small. 1972. *The wages of war.* New York: John Wiley.

Smith, H. 1970. U. S. assumes the Israelis have A-bomb or its parts. *New York Times* 18 July.

Snyder, Richard C., H. W. Bruck, and Burton Sapin eds. 1962. *Foreign policy decision making.* New York: Free Press.

Sommer, Theodore. The 1966.objective of Germany. In Buchan 1966, 39-55.

Spence, J. E. 1974. Republic of South Africa: Proliferation and the politics of outward movement. In Lawrence and Larus 1974, 209-33.

Spiegel, Steven L. 1972. *Dominance and diversity*. Boston: Little, Brown.

Spykman, Nicholas J. 1942. *America's strategy in world politics: The United States and the balance of power*. New York: Harcourt, Brace.

Steinbruner, John D. 1974. *The cybernetic theory of decision*. Princeton: Princeton University Press.

Stephenson, Richard. 1954. *Introduction to nuclear engineering*. New York: McGraw-Hill.

Stockholm International Peace Research Institute. 1972. *The near-nuclear countries and the NPT*. New York: Humanities Press.

———. 1974. *Nuclear proliferation problems*. Cambridge, Mass.: MIT Press.

———. 1977. *World armaments and disarmament: SIPRI Yearbook 1977*. Cambridge, Mass.: MIT Press.

Stoessinger, John G. 1969. *The might of nations: World politics in our time*. New York: Random House.

Stover, R. L., and G. K. Moeller. 1961. *Methods for determining fuel burnup*. Oak Ridge, Tenn.: MIT Engineering Practice School.

Subrahmanyam, K. 1972. The role of nuclear weapons: An Indian perspective. In Holst 1972a, 131-41.

———. 1974. India: Keeping the Option Open. In Lawrence and Larus 1974, 112-48.

Time. 1976. How Israel got the bomb. 12 April, 39-40.

Time. 1977. Uranium: The Israeli connection. 30 May, 31-34.

United Nations. 1950-82. *U. N. statistical yearbook*. Series. New York: United Nations.

———. 1968. *Effects of the possible use of nuclear weapons and the security and economic implications for states of the acquisition and further development of these weapons*. New York: United Nations.

U. S. Arms Control and Disarmament Agency. 1977. *Arms control and dis-armament agreements*. Washington, D. C.: U. S. Arms Control and Disarmament Agency.

U. S. Atomic Energy Commission. 1955. *Research reactors*. New York: McGraw-Hill.

———. 1956. *Seventeenth/eighteenth semi-annual report(s) of the AEC*. Washington, D. C.: U. S. Government Printing Office.

———. 1957. *Symposium on the reprocessing of irradiated fuels*. Vols. 1-3. TID-7534. Washington, D. C.: U. S. Government Printing Office.

———. 1968. *AEC gaseous diffusion plant operations*. ORO-658. Washington, D. C.: U. S. Government Printing Office.

———. 1971. *Reactor fuel cycle costs for nuclear power evaluation*. WASH-1099. Washington, D. C.: U. S. Government Printing Office.

———. 1974. *Nuclear power growth 1974-2000*. WASH-1139(74). Washington, D. C.: U. S. Government Printing Office.

U. S. Department of Commerce. 1965. *Supplement to the survey of current business*. Washington, D. C.: U. S. Department of Commerce.

U. S. House of Representatives. 1975. Nuclear proliferation: Future U. S. foreign policy implications. Hearings before the Subcommittee on International Security and Scientific Affairs of the Committee on International Relations, 94th Congress, 1st session, 21, 23, 28, 30, October and 4 and 5 November.

U. S. Senate. 1968. Non-proliferation treaty. Hearings before the Committee on Foreign Relations, 90th Congress, 2d session, 10-12 and 17, July.

———. 1976. Export reorganization act of 1976. Hearings before the Government Operations Committee, 94th Congress, 2d session, 19, 20, 29, 30, January and 9 March.

U. S. Senate, Committee on Government Operations. 1975. *Facts on nuclear proliferation*. Committee Print, 94th Congress, 2d session, December.

Van Cleave, William. 1974. Nuclear

technology and nuclear weapons. In Lawrence and Larus 1974, 30-68.

Vital, D. 1969. The problem of guarantees. In Barnaby 1969a, 144-53.

Wakaizumi, Kei. 1966. The problem for Japan. In Buchan 1966, 76-87.

Washington Post. 1980. Qaddafi's two tries to buy the bomb. 16 May, 121.

Washington Star. 1981a. Saudis reported aiding in bomb. 19 January, 13.

———. 1981b. Nigeria says sold uranium to Libya, 14 April, 5.

Weart, Spencer R. 1976. Scientists with a secret. *Physics Today* 29, no. 2 (February):23-31.

Weidenbaum, Murray. 1967. Defense expenditures and the domestic economy. In *Defense management*, ed. Stephen Enke. Englewood Cliffs, N. J.: Prentice-Hall.

Weiner, Charles. 1978. Retroactive Saber Rattling? *Bulletin of the Atomic Scientists* 34, no. 4 (April):10-13.

Wentz, Walter B. 1968. *Nuclear proliferation.* Washington, D. C.: Public Affairs Press.

Wick, O. J. 1967. *Plutonium handbook: A guide to technology.* New York: Gordon and Breach.

Wilkenfeld, Jonathan. 1972. Models for the analysis of foreign conflict behavior of states. In *Peace, war and numbers*, ed. Bruce Russett. Beverly Hills, Calif.: Sage Publications.

Wilkinson, W. D. 1962. *Uranium metallurgy.* Vol. 1. New York: Interscience.

Willrich, Mason. 1971. *Civil nuclear power and international security.* New York: Praeger.

Willrich, Mason, and Theodore Taylor. 1974. *Nuclear theft: Risks and safeguards.* Cambridge, Mass.: Ballinger.

Winkler, Robert L. 1972. *An introduction to Bayesian inference and decision. New York: Holt, Rinehart, and Winston.*

Winkler, Theodor. 1981. Israel's preventive strike. *International Defense Review 14, no. 7: 838.*

Wohlstetter, Albert, et al. 1975. Moving toward life in a nuclear armed crowd. Los Angeles: Pan Heuristics.

———. 1979. *Swords from plowshares.* (Chicago: University of Chicago Press.

Wolfe, Thomas W. 1970. *Soviet power and Europe.* Baltimore: Johns Hopkins Press.

Wolfers, Arnold. 1962. *Discord and collaboration.* Baltimore: Johns Hopkins Press.

Wood, David. 1968. *Conflict in the twentieth century.* London: International Institute for Strategic Studies.

Wordsworth, A. D. 1969. Factors affecting the cost of nuclear fuels and the selection of reactor fuel cycles in developing countries. In *Nuclear energy costs and economic development*, 317-38. Vienna: International Atomic Energy Agency.

Wu, Leneice N. 1972. The Baruch plan: U. S. diplomacy enters the nuclear age. Prepared for the Subcommittee on National Security Policy and Security Development of the Committee on Foreign Affairs, U. S. House of Representatives.

Wu, Yuan-li. 1970. *The organization and support of scientific research and development in mainland China.* New York: Praeger.

Yaeger, Joseph A., ed. 1980. *Nonproliferation and U. S. foreign policy.* Washington, D. C.: Brookings Institution.

Zoppo, Ciro. 1964. France as a nuclear power. In Rosecrance 1964a, 113-56.

Index

Index

Hot cells, 37, 38, 138
Hungary, 41

Incentives, 14, 46–49, 65–67. *See also* Motive factors
India, 41, 51, 53–54, 72, 73, 122–24, 127, 138; first proliferation decision, 6, 7–8, 170; second proliferations decision, 7–8, 27, 30–31, 171
Indicators, motivational, 44–46; technical, 35–38, 187–93. *See also* Nuclear infrastructure
Indonesia, 8
International Atomic Energy Agency (IAEA), 24, 71, 181, 183
International Nuclear Fuel Cycle Evaluation (INFCE), 4
Iran, 51, 138, 153, 158, 159
Iraq, 51, 136–39, 153, 158, 159–60, 212n.7; Osiraq, attacked by Israel, 6, 10, 26, 45, 72, 136–38, 148, 159
Isotope separation. *See* Enrichment
Israel, 7–8, 29, 41, 56, 117–18, 153, 154, 170–71. *See also* Iraq, Osiraq
Italy, 41, 51, 53, 153, 154; as supplier, 138, 159

Japan, Imperial, 6, 7, 8, 41, 51, 53, 79, 167; postwar, 41, 51, 53, 153, 154

Korea: North, 15, 153, 158; South, 6, 7–8, 14–15, 41, 56, 66, 69, 124–26, 127, 153, 156–57, 172

Lag time, 149–52, 153
Lamarsh, John, 180–81
Latent capacity, 5–6, 13, 31–35, 36; countries with, 40–43; defined, 1, 31, 37; longevity of, 75–77, 80; threat from (*see under* Motive factors). *See also* Capability decision
Libya, 7, 8, 70, 135, 139–41, 143, 153, 158, 159
Likelihood, 146–47. *See also* Latent capacity; Propensity
London Suppliers Group, 4, 15
Lovins, Amory, 10

McCone, John, 39
Marcoule, 23, 121
Metal conversion, 36, 180, 196–97
Mexico, 51, 153, 158
Milling, 36, 178, 195–96
Mining, 36, 179, 195
Motivationsl hypothesis, 12–16, 43–44, 82, 91–111. *See also* Propensity

Motive factors, 74, 96, 102; conventional threat, 60–63, 115, 117, 118, 119, 122, 125, 138; domestic turmoil, 63–64, 123, 132, 135; economic burden, 65, 138; global power, 51–55, 120–21, 123; latent threat, 59–60, 114–15, 119, 122, 135, 138, 139; nuclear ally, 55–56; nuclear threat, 56–59, 125; pariah, 55–56, 117, 118, 133; regional power, 50–51, 116, 120, 129, 139; regional proliferation, 64–65, 121, 122, 125, 133, 135, 138; war loss, 64, 121, 122, 132, 135, 138
Mozambique, 117

NATO, 107–8, 121
Nehru, Jawaharlal, 127, 170
Netherlands, 41, 153, 154
Nigeria, 51, 140, 153, 158
Nitric acid, 188–89
Nixon, Richard, 125
Non-Proliferation Treaty (NPT), 15, 24, 70, 126, 130, 133, 135, 138–39, 143
Norway, 41, 153, 156
NRX, 122
Nuclear infrastructure, 35–38, 83–87. *See also* Latent capacity
Nuclear weapons: minimum useful number, 20–25; production of, 19, 36–38, 177–84, 201–2
Null hypothesis (sui generis), 17–18, 82

OAS, 129
OAU, 116
Operational capability, 6, 14. *See also* Proliferation decision
Oppenheimer, J. Robert, 11
Osiraq. *See under* Iraq

Pakistan, 4, 6, 8, 64, 66, 134–36, 139–40, 153, 156–57
Park Chung Hee, 172
Peaceful Nuclear Explosives (PNE), 6, 20, 30–31, 170, 208n.12
Plutonium, 4, 20–22, 25, 36, 37, 174–77. *See also* Production reactors; Reprocessing
Poland, 41
Power reactors, 2–3, 37, 129, 135, 139
Predictors. *See* Dissuasive factors; Motive factors
Production reactors, 22, 23, 36–38, 121, 175–76, 180–82, 197–98
Proliferation decision, 5, 78, 81, 96. *See also under individual countries*

228